ANATOMICAL CASE HISTORIES

MOVING IN

Christy Hartman

I'LL NEVER FORGET MY FIRST BEE STING. ALL OF US KIDS WERE standing shoulder to shoulder on the side of the pool, waiting for the whistle to signal the end of break. Just as I jumped, the bee got trapped between my toes, freaked out and stung me. It seemed to happen all at once: the jumping and falling, the shrill whistle, the hot skin hitting cold water, and the burning, piercing pain. I remember the stinger was lodged between my toes and only came out when someone dug it out with a needle. That someone could only have been her, my mother.

When my mom tells me that the cancer in her breasts has spread to her lungs, we spend hours on the phone crushing on our dreams. Should we find one house, or two houses in the same town? We both value independence, but she wants to live close enough to

ANATOMICAL CASE HISTORIES

A PROBLEM – SOLVING TEXT IN ANATOMY

LAWRENCE K. SCHNEIDER, Ph.D.

Professor of Anatomy
Director, Division of Biomedical Sciences
School of Medical Sciences
University of Nevada
Reno, Nevada

YEAR BOOK MEDICAL PUBLISHERS, INC.
35 East Wacker Drive/Chicago

Printed in the United States of America.

Library of Congress Catalog Card Number: 75-44534

International Standard Book Number: 0-8151-7561-2

TO KAREN, BRANDT, KRISTIN AND ERICA

PREFACE

The idea for this book was conceived some years ago during a particularly heightened period of medical student questioning regarding the relevance of various aspects of their curriculum, anatomy not excluded. It became apparent to me that certain students were not achieving gratification from their studies in anatomy, perhaps because of long, tedious memorization sessions at the dissecting table or because of the absence of integrated functional or clinical material. The idea gained momentum when it became obvious that practicing physicians and dentists who returned periodically for continuing medical education voiced a common and serious concern that they "wished they had learned their anatomy at the time it had been available."

One specific experience convinced me that a book of this type might be very beneficial to the student who, for whatever reason, could not immediately recognize that a satisfactory knowledge of anatomy is essential to the successful practice of medicine. Medical students at this institution receive most of their anatomical instruction in their second year, *after* they have experienced an introduction to clinical medicine in the didactic curriculum and with physicians in the local community. In contrast to the students who frequently voice a desire to simply "get through with" their anatomy, these individuals approach the subject eagerly and are considerably more receptive to instruction. It is the aim of this book to demonstrate the value of a sound foundation in anatomy to the student while he or she is actively engaged in learning the subject.

Although this book deals with detailed discussion of the anatomical substrates for various clinical problems, it in no way is intended to serve as a textbook of clinical medicine. Thus, the only allusions to treatment appear in statements such as "surgery was performed," "antibiotics were prescribed" or "the bone was set and placed in a cast." All histories are fictitious, although they often are loosely based on actual occurrences described in the literature or on the experiences of various clinical associates of mine. Each history is followed by pertinent questions designed to encourage the student to diagnose the problem and to discuss the anatomical basis for the particular disorder. He may then compare his conclusions with those described here in the discussion of each case. The latter include varying degrees of detail sufficient to achieve the intended point or points. The student is encouraged to consult standard textbooks of anatomy where appropriate.

The text is arranged according to organ systems, beginning with the skin and musculoskeletal systems and proceeding through respiratory, cardiovascular, gastrointestinal, renal, head and neck, nervous, endocrine and reproductive. Certain chapters, notably those dealing with the musculoskeletal system and the head, neck and special senses, are considerably longer than others since disorders of these aspects of the body are most frequently encountered by the physician. The major emphasis is on gross anatomy, but significant aspects of histology, neuroanatomy and embryology are included. Although many aspects of human anatomy could not be incorporated in a book of this size, it is hoped that through its study the student will begin to think in terms of the anatomical basis for a number of additional clinical problems. In addition, the book should serve as a ready source for review by both medical students and physicians of fundamental anatomical principles relevant to disease and/or trauma.

I wish to express my gratitude to the numerous individuals who assisted in the preparation of this book, to Linda Forson Hall for her lucid illustrations and to Year Book Medical Publishers for their enthusiasm and support.

LAWRENCE K. SCHNEIDER

TABLE OF CONTENTS

Chapter 1

INTEGUMENTARY SYSTEM

CASE 1–1: LYMPHATIC DRAINAGE OF THE SKIN

History

A 39-year-old man noted over a period of several weeks that a mole on the left side of the anterior abdominal wall at the level of the umbilicus was darkening and becoming larger. He was prompted to visit his dermatologist when he observed a reddish circular inflammatory area around the mole. The physician removed the structure surgically, excising a considerable amount of surrounding skin and subcutaneous tissue. The structure was fixed and sent to the pathology laboratory, where it was confirmed as being malignant. In addition, the physician palpated the patient's axillary and inguinal lymph nodes on the left side and found them to be slightly enlarged and hardened. Needle biopsies of these nodes demonstrated the presence of metastatic processes and, therefore, block dissections and removal of the glands were performed. Both the left axillary and inguinal areas were treated heavily with x-rays following the operations. Twenty-seven months later there were no recurrences of the patient's disorder, and the prognosis for a technical "cure" was good.

Questions

From what condition was the patient suffering?

Why did his physician palpate the lymph nodes in both the axillary and inguinal regions?

What is the pattern of lymphatic drainage from the general area of the man's disorder?

Describe the positions and relationships of the various axillary and inguinal lymph nodes.

Trace the lymphatic drainage from both sets of nodes to the general venous system of the body.

Discussion

This condition, known as *malignant melanoma,* is defined as "a malignant tumor, usually developing from a nevus and consisting of black masses of cells with a marked tendency to metastasis."[1] Various sets of lymph nodes were palpated in this case because of the invasiveness of this type of tumor; similarly, during surgical excision of the tumor, considerable surrounding tissue was removed to ensure elimination of all cancerous cells. The abdominal wall superior to the umbilicus drains to the axillary lymph nodes, whereas the region of the wall inferior to the umbilicus drains to the inguinal nodes (Fig. 1–1). Since the tumor was located in the transition area, which may drain in either or both directions, both sets of lymph nodes were examined. As might be expected, both were found to contain metastatic tumor cells from the melanoma and, therefore, both were treated surgically as well as radiologically.

The *lymph nodes of the axilla* are distributed largely within the fat of this area and often lie in close association with blood vessels. They are arranged for descriptive purposes into five main groups as follows: (1) The *pectoral nodes,* three to five in number, lie along the lateral border of the pectoralis minor muscle in close relationship to the lateral thoracic artery and veins. They drain the anterior and lateral regions of the thoracic wall and the majority of the ipsilateral breast. (2) The *lateral nodes,* also three to five in number, lie largely behind (sometimes medial to) the axillary vein. They constitute the primary draining nodes of the upper limb. (3) The *subscapular (posterior) nodes,* usually five or six in number, occupy a position deep in the axilla along the subscapular blood vessels near the lateral border of the scapula. They drain the posterior portion of the thoracic wall and shoulder area. (4) The *central nodes* are embedded in fat near the base of the axilla. These nodes usually number 10 to 12, and they may be large. It is this group that is most frequently palpable. They receive afferent vessels from the preceding three sets of nodes and send efferents to the apical nodes. (5) The *apical group,* described variously as possessing from one to 14 nodes, lies medial to

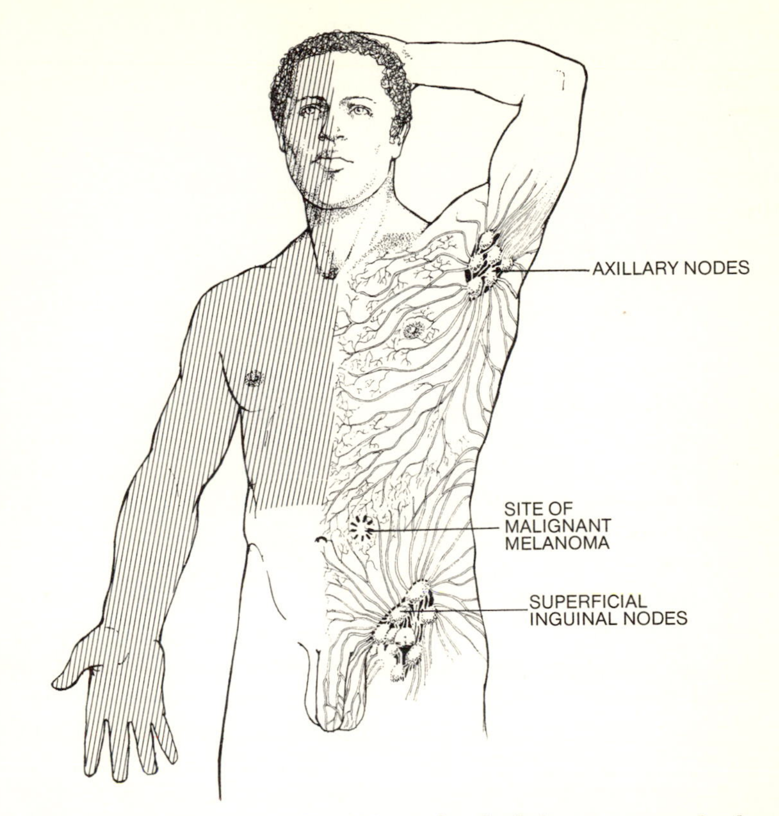

Fig. 1–1.–The left side of the individual demonstrates the lymphatic drainage of the anterior thoracic and abdominal walls. Note that above the umbilical level, drainage is to *axillary nodes*, whereas below it drainage is to *superficial inguinal nodes*. The *shaded quadrant* is the only portion of the body not drained by the thoracic duct. (Based partially on Sappey.)

the pectoralis minor muscle in contrast to the previous groups, which lie lateral to it. This group occupies a position behind the clavipectoral fascia in the apex of the axilla. These nodes receive afferents from all of the other groups as well as occasionally some direct branches from the breast. They drain via the *subclavian trunk* (sometimes multiple) to the systemic venous circulation (see below).

The *inguinal lymph nodes* are located just below the inguinal ligament in the upper portion of the femoral triangle. They are arranged in two groups, the superficial and deep inguinal nodes. The *superficial inguinal lymph nodes* also may be subdivided into two sets, which together form a T in the anatomical position. The horizontal limb or set roughly parallels the inguinal ligament and consists of about five or six nodes. The lateral nodes in this set receive lymph from the gluteal region and from the anterolateral, lateral and posterolateral regions of the abdominal wall below the level of the umbilicus. The medial nodes of this set receive lymph from the skin around the anus, from the external genitalia (see also Case 10–1) and from the anteromedial portion of the abdominal wall below the umbilicus. A few lymphatic vessels from the body and/or fundus of the uterus also may reach these nodes via vessels that accompany the round ligament through the inguinal canal. The vertical set of superficial nodes (four or five in number) lie along the great saphenous vein and receive, with the exception of a portion of the calf, all of the lymph from the lower extremity. The *deep inguinal nodes* usually number only one to three. They are situated deep to the fascia lata at the medial side of the femoral vein. One of the nodes is frequently found with the fat occupying the femoral canal. These nodes receive afferents that accompany the femoral vessels, the lymphatics from the glans penis or clitoridis and some from the superficial inguinal nodes. Note that the testis does not drain to the inguinal lymph nodes but, rather, to internal lymph nodes (see also Case 10–1).

The axillary lymph nodes drain via the subclavian trunks in a variety of manners. The right subclavian trunk may enter the venous system directly, entering either the internal jugular vein, the subclavian vein or the point where the two join to form the right brachiocephalic vein. Or it may enter the formation of the *right main lymphatic duct,* which arises from the junction of the right internal jugular trunk, subclavian trunk and bronchomediastinal trunk. The left subclavian trunk may enter either the venous system, as was the case on the right side or, as is more often the case, the *thoracic duct,* which in turn empties into the venous system on the left. The inguinal lymph nodes drain to *external iliac nodes,* then to *common il-*

iac nodes and eventually to *lumbar (aortic) nodes*. The efferent vessels from the right and left lumbar nodes join to form the *right* and *left lumbar trunks*, respectively, which contribute to the formation of the *cisterna chyli*. The thoracic duct leads from the cisterna chyli through the diaphragm, ascends through the posterior and superior mediastina and into the left side of the root of the neck. Here it most commonly empties into the junction of the left internal jugular vein with the left subclavian vein (i.e., at the site of formation of the left brachiocephalic vein), although it may enter either vessel prior to their union. Thus, the thoracic duct ultimately drains all of the body except the upper right quadrant, which is drained by the right main lymphatic duct or by its tributaries in the event that the main duct is not present as a specific entity (see Fig. 1–1).

CASE 1–2: HISTOLOGY OF THE SKIN

History

A 27-year-old man had a growth with characteristics almost identical to the one described in Case 1–1 (a darkly pigmented spot surrounded by a halo of inflamed skin). He possessed a very fair complexion with numerous freckles and blond hair. He spent a great deal of time in the sun and sustained frequent sunburns. The pigmented spot was removed surgically and diagnosed in the pathology laboratory as an early malignant melanoma. There was no evidence of metastasis and, as a result, the physician was optimistic about the patient's chances of survival. He was admonished, however, to stay out of the sun as much as possible.

Questions

Describe the normal histology of the epidermis.

Where do melanocytes originate and where do they reside after migration?

Describe the differences in metabolism of these cells that occur in health and in their reactions to ultraviolet radiation and disease (malignancy).

What are the major criteria for determining malignancy in melanotic tumors?

Discussion

The *epidermis* does not possess a direct arterial blood supply but, rather, is nourished by diffusion of substances from the underlying dermis. The epidermis may be subdivided for descriptive purposes and/or according to function into four layers. It is important to note that only in the palms of the hands and soles of the feet (i.e., where the epidermis is thickest) are all the layers clearly present. In most parts of the body, only the innermost and outermost layers may be distinguished.

The innermost layer is known as the *stratum germinativum.* There is a single layer of columnar *basal cells,* which abuts on the basement membrane separating them from the dermis. The adjacent cells of the stratum are often referred to as *prickle cells* because they appear to be joined by cellular spinous processes, which contain *tonofibrils* that end on *desmosomes.* Mitoses are found in these two cell types of the stratum germinativum (hence its name). From these, daughter cells migrate toward the surface, pushing other cells in front of them.

The next layer from inside out is the *stratum granulosum,* so named because the component cells contain *keratohyalin granules.* The cells have become flattened, they stain more deeply than the preceding layer and their nuclei show degenerative changes. Adjacent to this layer is the *stratum lucidum,* so named because of its transparent appearance. The cells demonstrate increased degeneration and are either dead or dying at this point. Nuclei and cytoplasmic organelles are no longer evident.

The surface layer is known as the *stratum corneum* because its cells are cornified or keratinized. This layer is especially thick in places of wear, i.e., the palms and soles. The cellular elements are flattened and densely stacked. The surfaces demonstrate interdigitations held together by desmosomes. The cytoplasm has been replaced largely by amorphous keratohyalin and remnants of tonofibrils. The cells at the surface are continually being sloughed off to be replaced by cells from the deeper layers.

Skin is pigmented largely by the amount of *melanin* it contains. White people have a minimum of this black pigment and, therefore, their skin color is a combination of the basic yellow caused by carotenes and the pink caused by the presence of blood vessels in the dermis. In blacks, on the other hand, melanin is plentiful, and not only in the basal layer; it may extend up into the stratum granulosum. Moles, freckles, areolae of the nipples and other pigmented areas of the body (i.e., perianal skin) contain increased concentrations of melanin. *Melanocytes* are the cells that contain the pigment. They arise from neurectoderm of the neural crests as stem cells, the *melanoblasts,* which migrate through the dermis during fetal life to arrive at the dermoepidermal junction, i.e., the basal cell layer. Pigment is contained within these cells in *melanosomes.* Ultraviolet radiation initially causes a darkening of already-present melanin, followed by new melanin synthesis. This excess pigment eventually degenerates as *melanoid* unless there is continuous or repeated exposure to sunlight.

According to Pinkus and Mehregan[2] the criteria for diagnosing malignancy of melanotic tumors, in which moles or other pigmented spots become cancerous, are (1) the presence of mitotic figures, (2) an inflammatory reaction composed of lymphocytes and perhaps also plasma cells and (3) origin at the dermoepidermal junction.

CASE 1-3: CLEAVAGE LINES OF THE SKIN

History

A 60-year-old man had had numerous sebaceous cysts removed from the back of the neck as a young man. The work was performed by an old country doctor who did not use anesthesia or aseptic conditions. Most of the cysts had been opened by incisions made with a straight razor, the lines of the incisions running in all directions. Some of the cysts had become infected. As a result of these rudimentary operative procedures, the man's neck was pock-marked, scarred and unsightly. The scars that had a vertical orientation were strikingly more pronounced and severe than those running in the horizontal plane. Due to his relatively advanced age, the patient decided against plastic surgery designed to remove or reduce the scarring.

Questions

What accounted for the fact that the scars that were oriented vertically were more severe than those with a horizontal orientation?

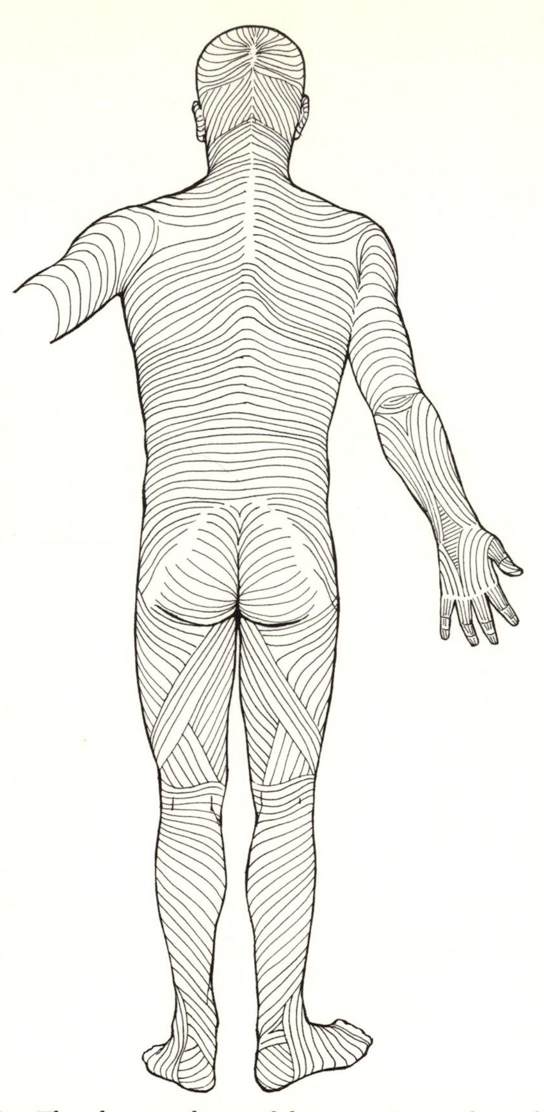

Fig. 1–2.—The cleavage lines of the posterior surface of the body. Incisions made parallel to these lines heal with a fine scar, whereas those made perpendicular often result in gaping wounds and unsightly scars. (Based on Gray, H.: *Anatomy of the Human Body.* Goss, C. M., ed. [29th Am. ed.; Philadelphia: Lea & Febiger, 1973].)

Are there places on the body (other than the obvious lines of the palms) where some sort of linearity is apparent?

Discussion

If a sharp conical instrument pierces the skin and is then retracted, rather than leaving a circular wound, it leaves a slit. If this is done over the entire surface of a cadaver, a pattern of these *cleavage lines (tension lines, Langer's lines)* becomes apparent. Collagenous fibers in the skin and subcutaneous tissue tend to run parallel to these lines. These lines are related to a certain degree to the crease lines found in the various parts of the body, but they also are found in areas where the latter are not evident. Although the pattern of these cleavage lines is not the same for all individuals, there is a fair amount of consistency in their distribution. The pattern of these lines on the back of the body is presented in Fig. 1–2. There are only a few places on the body where this pattern seems to break down.

The significance of these lines lies in the observation that incisions made perpendicular to them frequently result in gaping wounds that tend to heal with unsightly scars. On the other hand, incisions made parallel to the orientation of the lines usually result in a fine scar that is much less unsightly than the former. As a result, surgeons are encouraged to commit to memory the patterns of these cleavage lines so as to minimize the production of unsightly scars. One place where these lines are particularly evident on most individuals is on the palmar surfaces of the middle digits of the fingers where they run in a plane parallel to the long axis of the fingers.

CASE 1–4: DERMATOMES

History

A 77-year-old man complained to his dermatologist of a unilateral band of painful vesicles and pustules extending around the abdominal wall from the umbilicus to the back on the right side. Examination revealed that this band ranged from about 1½–2½ inches in width. During the history the patient recalled that he had recently slipped it behind her ear. "I've worked with the parents in this town for thirty years. They are not progressive enough for the program you have in mind. Especially now."

Defiance flared up in me, unexpected and almost wholly unfamiliar. "I will not cancel this program."

Bernice looked at me over her glasses. It was like she was seeing through my momentary dissent, into the person she knew I really was.

"You'll do what you think is best, obviously. You're in charge."

When I walked back to my office and closed the door, I put my head down in my hands.

Sitting in my office, I read the news that accrued over the next few days. As I read the horrible details of the crime, I kept thinking about Horace Wilbur's paintbrushes. The little ones in little hands that he had covered with his own, larger hand, and guided through the delicate strokes it took to create a bird feather. Had there been a packet of them in his car that day, the day he pointed a gun at the teenager's face, and pulled the trigger? Had they rolled out onto the ground when he shoved her body out the door, rolled and dragged it into a field, doused it with gasoline? Had he ever told her the importance of the details in creating reality, as I'd heard him tell so many children?

I dialed Quinn's phone number.

"I have some bad news," I said, the tremor back in my voice. "We've cancelled your program for the summer. We're hoping that next year, things will have calmed down. We don't want to lose you as a volunteer."

I rambled on, though Quinn seemed to have resigned himself to what I was saying some time before. Had he heard similar things in the past? Was anything coming out of my mouth new to him? Likely not. I spoke of all the wonderful plans we could move forward with next summer, when things had gone back to normal, when Horace Wilbur was in prison, and the world was a little better.

been exposed to chickenpox by one of his great grandchildren. He remembered that he also had had this disease as a child. The current problem had begun as a skin rash manifested by erythematous maculopapular plaques, which later became vesicular and then pustular, finally becoming crusted 2 or 3 weeks after onset. The patient was given idoxuridine to be applied topically, which seemed to relieve the symptoms to some degree. The condition eventually disappeared, probably independent of the treatment.

Questions

What is the name of this condition?

What relationship does it have to chickenpox and other diseases or conditions?

What part of the skin does it affect?

Why does it usually only affect a band of skin, as in this case?

Would you expect to observe the same bandlike arrangement of the vesicles in the limbs as is seen in the trunk?

Discussion

This disease is known as *herpes zoster* or "*shingles,*" and it is not uncommon in individuals over 50 years of age. It has a higher incidence in people who have had chickenpox (varicella) as children. These two clinical entities are caused by the same virus, which apparently lies dormant in the adult until reactivated by such situations as reexposure (as, for example, to a child with chickenpox) or by x-irradiation. Zoster also is more prevalent in individuals with Hodgkin's disease or leukemia.

The disease is characterized by the intranuclear location of the virus and by the specific epidermal reaction. The latter is manifested by edematous cells or groups of cells floating in the vesicles like balloons. These epithelial cells may be multinucleated. Cell death occurs in the so-called *malpighian layer* of the epidermis. This is the area that contains the prickle or spinous cell type (see Case 1–2). In serious cases of the disease, nearly total destruction of parts of the epidermis may occur.

It appears that after infection by the virus in childhood

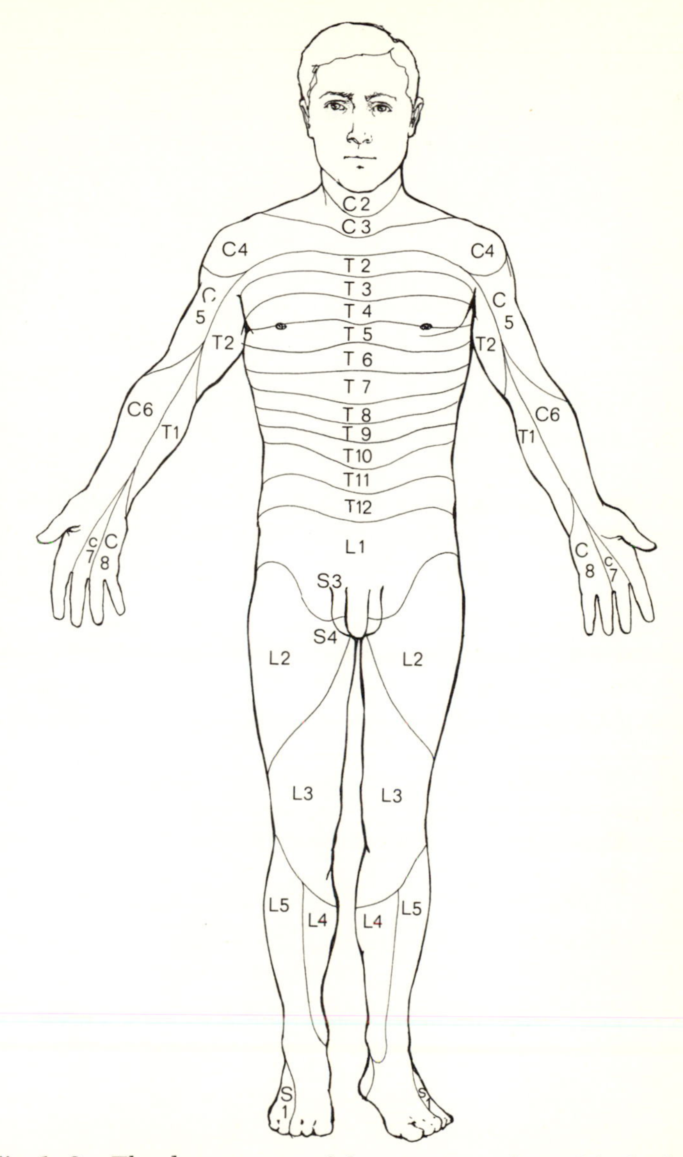

Fig. 1–3.—The dermatomes of the anterior surface of the body. In the trunk these appear as bandlike segments, whereas in the limbs the pattern is less regular. (Redrawn from Barr.[3])

chickenpox, the virus may lie dormant for many years in one or more *dorsal root ganglia* following migration to this structure via a sensory nerve. Following reexposure or reactivation by x-rays, the virus appears to migrate down the sensory nerves to the skin subserved by that dorsal root ganglion. The area of skin thus affected delineates a *dermatome,* or the bandlike portion supplied by that single ganglion. Indeed, this disease facilitated our knowledge of dermatones; neurologic investigations of these complexes have been frustrated by the ubiquitous overlap of cutaneous sensory nerves. Destruction of a single dorsal root or its ganglion does not produce complete anesthesia but, rather, hypoesthesia, or reduced cutaneous sensation.

The pattern of dermatomes on the anterior surface of the body is presented in Figure 1–3. Note that the bandlike arrangement of dermatomes is characteristic of the trunk but becomes more confused in the limbs. In the embryogenesis of the limbs, the central segments of the limb buds grow out distally, carrying those nerve fibers, whereas the peripheral areas grow more slowly. Thus, in the upper limb, segments C_5 and T_2 are found in close apposition in the proximal portion of the arm, whereas segments C_7 and C_8 extend down into the forearm and hand. The dorsal root ganglion that was affected in this case was T_{10} on the right side. This ganglion subserves the sensory fibers that extend in a band around the trunk at about the level of the umbilicus.

CASE 1–5: CUTANEOUS INNERVATION

History

For an abdominal exploratory operation, the surgeon made a long vertical incision in the midsagittal plane of the patient's anterior abdominal wall. There was a nominal amount of bleeding in this procedure but, more importantly, upon healing of the incision, the patient experienced almost no loss of cutaneous sensation in the vicinity of the scar and no abdominal muscle paralysis. Since the patient was an elderly male, cosmetic considerations were not foremost in the surgeon's mind.

Questions

Describe very briefly the anatomy of the anterior abdominal wall as it applies to the incision made by the surgeon.

Describe the course and distribution of a spinal nerve with specific attention to the anterior abdominal wall.

Why did the surgeon choose the midsagittal plane for the incision rather than one that was parallel but more lateral in position?

Discussion

The two *rectus abdominis muscles* extend from the lower anterior aspect of the rib cage to the pubis. They are crossed by three or four *tendinous inscriptions* that divide the muscles into segments; these are apparent as individual bulges on the anterior abdominal walls of well-muscled male subjects. The rectus muscles are enclosed within the aponeurotic *rectus sheath* (Fig. 1–4), which prevents "bowstringing" of the muscles in such maneuvers as sit-ups. In general terms, the anterior lamella of the sheath is made up of the *external oblique aponeurosis* and half of the *internal oblique aponeurosis*. The posterior lamella above a point known as the *arcuate line* is composed of the other half of the internal oblique aponeurosis and the entire *transversus abdominis aponeurosis*. Inferior to the arcuate line, all layers pass in front of the muscles, leaving only the *transversalis fascia* behind the rectus muscles. A curved line may be observed on the surface at the lateral border of each rectus muscle; this is the *linea semilunaris*. The point at which the fibers from the right and left aspects of the rectus sheath interdigitate in the midsagittal plane (i.e., the line of the incision made by the surgeon) is known as the *linea alba* ("white line").

Each spinal nerve gives off a *meningeal branch* and then divides into an *anterior (ventral)* and a *posterior (dorsal) ramus*. The posterior rami pierce and supply the deep muscles of the back and then divide into *medial* and *lateral cutaneous branches*, which supply the skin of the posterior surface of the trunk. The anterior rami of the upper six thoracic spinal nerves

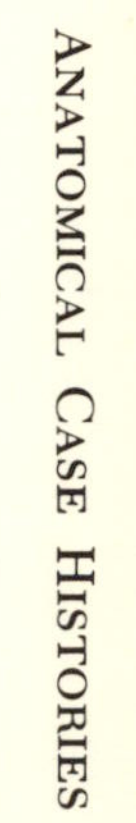

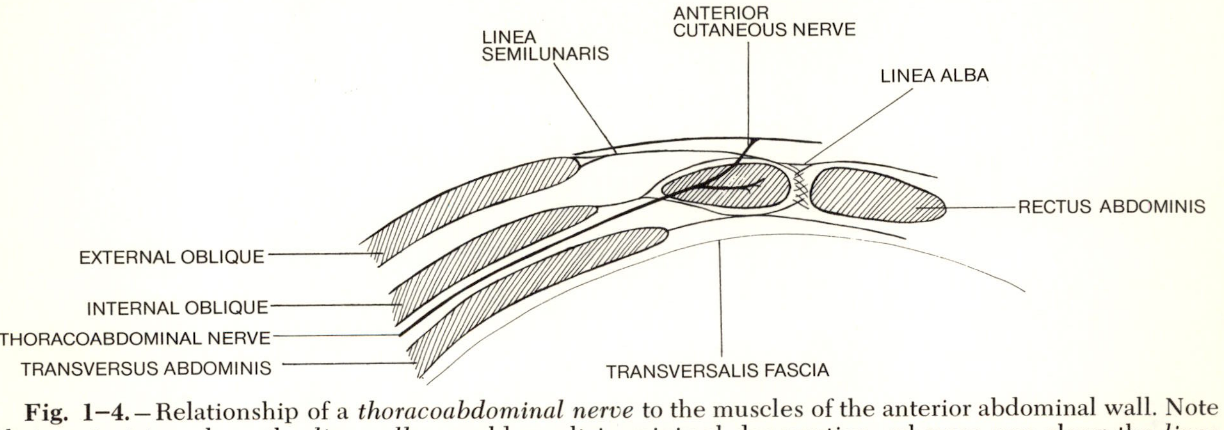

Fig. 1–4.—Relationship of a *thoracoabdominal nerve* to the muscles of the anterior abdominal wall. Note that an incision along the *linea alba* would result in minimal denervation, whereas one along the *linea semilunaris* would denervate the entire *rectus abdominis* muscle. An incision through the midline of the rectus would denervate its medial half.

are known as *intercostal nerves.* T_7 through T_{11} are intercostal posteriorly but, as they reach the midaxillary line, they begin to course inferiorly and medially to supply the anterolateral abdominal wall. They are therefore sometimes referred to as the *thoracoabdominal nerves.* The twelfth thoracic nerve lies below the last rib and is therefore referred to as the *subcostal nerve.* The thoracoabdominal nerves occupy the plane between the internal oblique and transversus abdominis muscles. In the midaxillary line they give off a *lateral cutaneous branch,* which divides into anterior and posterior branches. The nerve then enters the posterior lamella of the rectus sheath at the lateral edge of the muscle and divides into a small branch that supplies the muscle and a larger *anterior cutaneous branch* (see Fig. 1–4). The latter pierces the muscle, supplies it and then divides into *medial* and *lateral cutaneous branches* for supply of the skin of the anterior abdominal wall.

The surgeon chose the linea alba for his incision, since in that plane a minimum of nerve trunks would be cut (although there is considerable overlap of medial branches of anterior cutaneous nerves across the midline). An incision along the linea semilunaris (lateral border of the rectus muscle) would denervate the entire muscle, whereas one down the midline of the muscle would denervate the medial half of the muscle. These two approaches also would result in a much greater loss of cutaneous sensation on the anterior abdominal wall.

REFERENCES

1. Dorland: *Illustrated Medical Dictionary* (24th ed.; Philadelphia/London: W. B. Saunders Co., 1965).
2. Pinkus, H., and Mehregan, A. H.: *A Guide to Dermatohistopathology* (New York: Appleton-Century-Crofts, 1969).
3. Barr, M. L.: *The Human Nervous System* (2d ed.; Hagerstown, Md.: Harper & Row, 1974).

Chapter 2

MUSCULOSKELETAL SYSTEM

CASE 2–1: INNERVATION OF THE HAND

History

Between games of a state high school basketball tournament, the young star of one of the teams occupied himself by lying in his bunk and playing mumblety-peg with his pocket knife on the wooden floor. On one particularly difficult flip, the tip of the knife blade shallowly penetrated the thenar eminence of his right hand and then fell to the ground. The small cut seemed to require only superficial first aid treatment. The young man noticed, however, within a short period of time that he could not move his thumb as easily as usual and that the skin surrounding the knife wound had become numb. Specifically, he had lost many of the fine movements of the right thumb and could no longer touch the palmar surface of the tip of the thumb to the palmar surfaces of the tips of the other fingers. During the ensuing pregame warm up, the young player found that he could not handle fast passes (the ball tended to slide off the injured thumb), and (the injury being on his shooting hand) it quickly became evident that he had lost control of this motion. In effect, his game was so far off as the result of this seemingly minor injury that he had to sit out the remaining contest, and his team lost the championship. He was examined by a neurologist who offered a favorable prognosis with regard to recovery of the use of the thumb. This was indeed the case, and the young man was back in the starring role on the basketball team the following year.

Questions

What specific structure was severed by the tip of the knife?

From which segments of the spinal cord do these fibers arise?

What muscles were denervated as the result of this injury? Review their attachments to and actions on the thumb.

Why were some of these actions not totally lost but, rather, noticeable only at the level of control of fine movements?

Do any of these muscles possess dual innervation (i.e., from two major nerves)?

How do you explain the ballplayer's recovery from the apparently serious deficit?

Discussion

The structure that was severed in this case was a *branch of the median nerve* (sometimes referred to as the *recurrent branch*). This nerve supplies three of the intrinsic muscles of the thumb (the abductor pollicis brevis, the flexor pollicis brevis and the opponens pollicis), which form the majority of the bulk of the thenar eminence. The median nerve (Fig. 2–1) enters the hand by passing through the carpal canal under the

Fig. 2–1.—Relationship of the *median nerve* to the *flexor retinaculum* and *thenar muscles.* Section of the superficially placed *recurrent branch* results in considerable loss of thumb coordination and mobility.

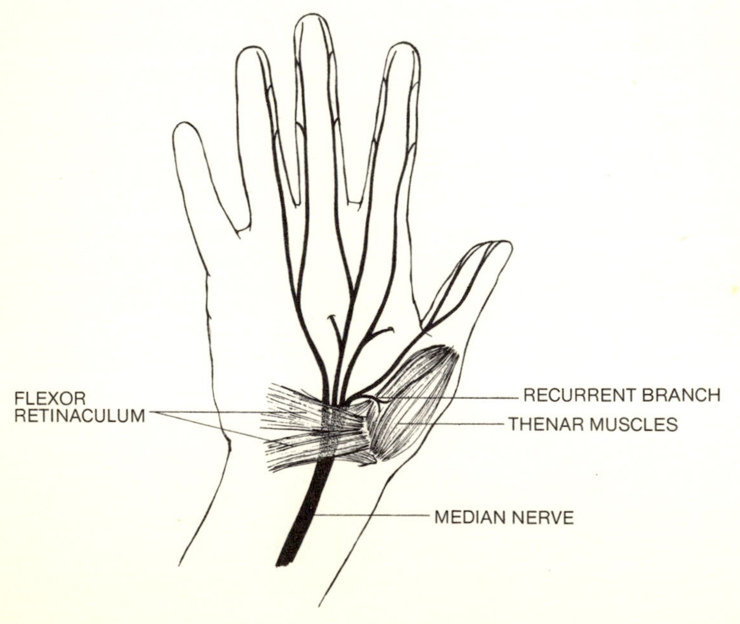

flexor retinaculum. At this point it lies between the flexor carpi radialis tendon laterally and the palmaris longus tendon medially (when the latter muscle is present). Upon emerging from the distal margin of the flexor retinaculum, the median nerve separates into medial and lateral divisions for eventual motor supply to most of the muscles of the thumb (see below) and the first two lumbricals, and for sensory innervation to the palmar surfaces of the lateral three-and-one-half fingers (and to the dorsal surfaces of the tip of some of these fingers). The remainder of the intrinsic muscles of the hand are supplied by the ulnar nerve (the student should review also the cutaneous innervation of the hand supplied by the ulnar and radial nerves). The first branch of the lateral division of the median nerve is the very superficial recurrent branch, which dives almost immediately into the muscles of the thenar eminence. The fibers in this nerve branch are derived from cervical spinal cord levels six and seven. It will be recalled that the anterior primary rami of spinal nerves C_6 and C_7 contribute to the upper and middle trunks of the brachial plexus, respectively. The anterior divisions of these two trunks form the lateral cord, which in turn contributes to the formation of the median nerve.

The *abductor pollicis brevis* is the most superficial and most laterally placed of the thenar muscles (see Fig. 2–1). It is a thin, flat muscle that originates from the anterior surface of the flexor retinaculum and from the scaphoid and trapezium bones. It is inserted into the lateral side of the base of the first phalanx of the thumb. The *flexor pollicis brevis* arises from the flexor retinaculum and trapezium medial to the abductor and inserts into the base of the proximal phalanx of the thumb and lateral sesamoid bone just medial to the insertion of the abductor. The flexor is frequently described as possessing deep and superficial heads. The superficial head, described above, occupies a plane similar, only medial, to that of the abductor. The deep head is very small and arises in close association with the oblique head of the *adductor pollicis* but inserts on the lateral sesamoid with the flexor. The nerve supply to the deep head of the flexor is the same as that of the adductor pollicis (ulnar), whereas the superficial head and the abductor are

Ohio, and the second highest rate of poverty. We could do something about it, we could make a difference. The idea of such a program was scintillating to me. I knew that I did good work, but to actually be in the field instead of trapped behind a desk making policies was something I'd dreamed about.

It hadn't mattered at the time that Quinn was transgender. But as I looked over the emails, the unspoken voices began to come back in the form of parents who could not understand such a thing.

"Hello," I said, when he answered his phone. "This is Mary Sneider from the library. I wanted to touch base about a few things. I assume you've heard the news?"

He was understanding; having worked with children before, he'd gone through extensive background checks. Of course he didn't mind having another one done. I felt relieved when I put the phone down. I turned to my computer and pulled up the flier we'd been making for our summer reading and writing program. "Build a Better World at the Lacon County Library," it said.

The children's librarian, Bernice, is in her sixties. She's been at the library for decades. I am technically her supervisor, but no one is under the delusion that she answers to anyone after her extended tenure. She had Quinn's website pulled up on the computer at her desk. It always gives me a mild shock to think about how proficient she is on the Internet. How she found this website long before I even thought to look for it. Listed were a series of stories, essays, and articles written by Quinn. Some were fictional worlds, and others were about struggles with his own mental health But many of them were about gay and transgender topics.

"The parents are not going to like it if they find this," Bernice said.

"It's a writing workshop for children," I said. My voice was quivering just a bit. It always does when I struggle with something. I tell myself that no one notices, but I am quite sure they all do. "It's on how to structure a short story."

"You know what I mean," Bernice said, readjusting her wire-frame glasses. An errant hair had escaped her tight grey ponytail, and she supplied by the median nerve. These paradoxes have led some authors to refer to the deep head of the flexor as the first palmar interosseous muscle. The *opponens pollicis* is situated deep to the abductor and the superficial head of the flexor. It similarly arises from the flexor retinaculum and trapezium, but it inserts on the lateral side of the entire length of the metacarpal bone of the thumb. It gives the impression of being wrapped around the metacarpal to some degree. This muscle, like the abductor and superficial head of the flexor, is supplied by the recurrent branch of the median nerve.

The actions of these muscles on the thumb are interdigitated and complex. It should be pointed out that, because the thumb is set at roughly right angles to the palm, the actions of these muscles are defined in relationship to the plane of the palm. Thus, a movement of the thumb away from the palm is termed abduction, whereas a return toward the palm is adduction (i.e., movements at right angles to the palm). Flexion of the thumb is a movement across (medially) and parallel to the palm, whereas extension is the movement in the opposite direction (lateral). The names of the abductor and flexor pollicis brevis muscles define what their actions are. Opposition of the thumb (the touching of the palmar surface of the thumb to the palmar surface of the other fingers) is a more complex movement and involves action of the abductor pollicis brevis and flexor pollicis brevis in addition to that of the opponens pollicis. The touching of portions of thumb and fingers other than the palmar surfaces is termed apposition.

Paralysis of the opponens is the most serious loss resulting from this type of injury, since the pincer grip between the thumb and index finger is critical to anyone who uses his hands for delicate work. Indeed, we have seen that it is probably indispensable to even less delicate arts such as catching a fast-thrown basketball. Control of thumb movement was not totally lost in this patient, however, since the innervation to the abductor pollicis longus and flexor pollicis longus muscles was still intact. Rather, we can think of this case as a loss of the "fine tuning" adjustment of the thumb. Since there is no "longus" counterpart for the opponens pollicis, denervation of this muscle resulted in inability to perform that function. Although the flexor pollicis brevis does receive a dual innervation, the deep head (which remained intact in this case) was apparently too small to counteract the loss of function of the superficial head of this muscle. Because the wound was superficial and there was no displacement of the two cut ends of the nerve, regeneration of the nerve fibers was facilitated, and the young man eventually recovered complete use of the thumb.

CASE 2-2: ELBOW STRAIN

History

A 17-year-old male member of the local high school tennis team developed pain and tenderness in the upper lateral aspect of the right forearm following a strenuous practice session early in the season. Specifically, the area surrounding and just distal to the lateral epicondyle of the humerus was the most sensitive to manipulation or pressure. Also, the pain was magnified during backhand shots but was minimal to nonexistent in the forehand stroke. The team trainer instructed the boy to use a tight elastic band around the upper forearm during play. This helped somewhat initially, but the pain continued and became chronic, severely hampering his ability to compete effectively as a member of the team. He was sent to a local orthopedic surgeon, who examined the arm, and to a radiologist, who ordered a series of films taken of the affected area. The latter were negative with regard to muscle, bone or joint injury or pathology. The boy was instructed to give up tennis until the condition cleared up; he was able to return to active competition after a 6-week layoff. With some slight correction in his technic, the condition did not recur.

Questions

What are the scientific and lay terms for this disorder?

What specific action of what portion of the upper limb (as might be performed in hitting a tennis ball) is responsible for this type of injury?

Are there other activities that would similarly affect this or the medial epicondylar area of the upper limb?

Does age have any affect on the severity of this condition?

Briefly review the positions and actions of the muscles of the forearm that are significant in the production of these symptoms.

Discussion

The scientific or medical term for this disorder is *lateral epicondylitis* and the lay term is "tennis elbow." Its usual cause is strenuous and repeated *extension of the wrist* against some force as might occur during the backhand stroke in tennis. This is the result of the fact that the majority of the extensors of the wrist and fingers originate from a common tendon on the *lateral epicondyle of the humerus* (Fig. 2–2). Contraction of these muscles against an oppositely directed force may cause a strain on the tendon and muscle fibers in the area around the lateral epicondyle, resulting in the pain and tenderness observed in this patient. The actual reason for the pain, whether it be in the muscle tissue or the tendon, or whether it be the result of tension on the periosteum of the lateral epicondyle itself, is unknown. An elastic band placed around the upper aspect of the forearm affords some protection against muscle strain, especially if it is used before any evidence of lateral epicondylitis is apparent. Once the condition has progressed to the point that it did in this case, the best treatment is probably rest, i.e., elimination of the aggravating factor. Keeping the elbow straight and not breaking the wrist on backhand shots tends to greatly reduce the strain on the lateral epicondyle and usually will prevent the development of tennis elbow.

Other activities, such as throwing a baseball, also can create a strain on the epicondylar areas, especially in young children in whom the secondary ossification centers for these areas have not yet fused to the main body of the humerus. According to Marks,[2] the ossification center for the medial epicondyle appears at about the fifth year of age and joins the shaft of the humerus around the eighteenth year; that for the lateral epicondyle appears at the twelfth year. At about age 14, these and other ossification centers (i.e., for the capitulum and trochlea) join to form a single epiphysis, which then joins the humeral shaft at between 16 and 17 years of age. Considerable controversy exists as to whether children below the ages at which these secondary centers have fused to the main body of the humerus should be encouraged to repeatedly perform acts

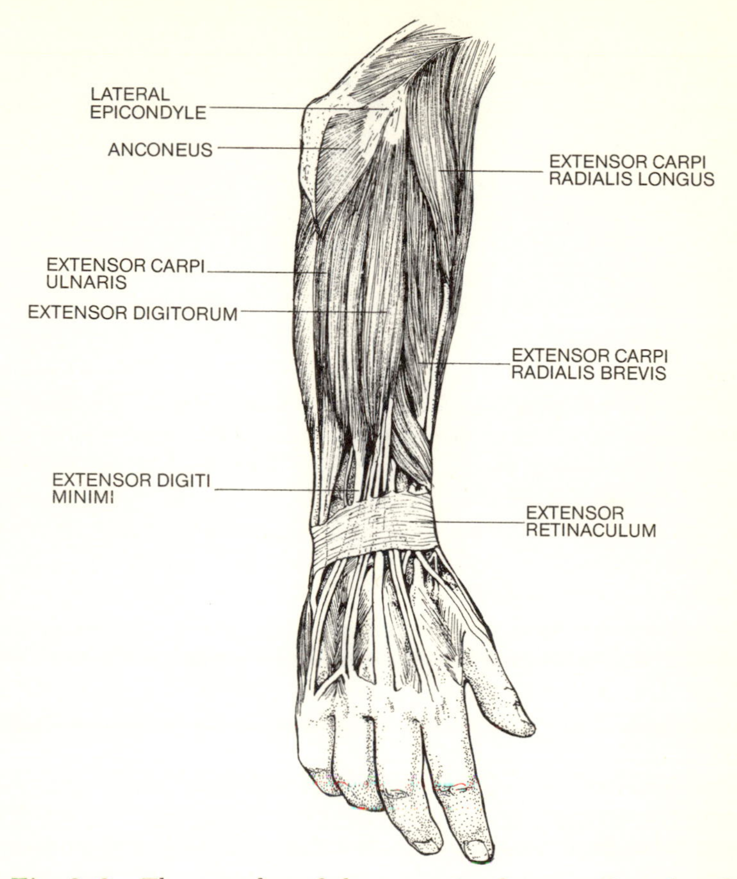

Fig. 2–2.—The muscles of the posterior forearm. Note that the majority of the extensors of the wrist and fingers originate from a common tendon from the *lateral epicondyle* of the humerus. (Redrawn from Tobin.[1])

that place great strain on temporary joints. "Little League elbow" frequently affects baseball pitchers in the pre-teen or early teen-age group. This is not to be confused with actual fracture of an epicondyle, which usually occurs as a result of direct trauma or violence to the elbow area rather than from muscle action.

The lateral epicondyle gives rise to many of the muscles on

the posterior forearm that extend the wrist and fingers and supinate the hand. These muscles also take partial origin from the thickened deep fascia that overlies them. One of the muscles, the *anconeus*, extends only to the superior portion of the ulna; it assists the triceps in extending the forearm. The remaining muscles are as follows: (1) The *extensor carpi radialis brevis*, which inserts on the back of the bases of the second and third metacarpals. (2) The *extensor digitorum*, which divides into four tendons, one for each finger except the thumb. Each tendon divides into a central slip, which inserts on the back of the base of the middle phalanx, and two collateral bands, which insert on the distal phalanx. The tendons expand medially and laterally to form the *extensor expansion* or *hood*. (3) The *extensor digiti minimi*, whose tendon joins that of the extensor digitorum to the little finger. (4) The *extensor carpi ulnaris*, which inserts on the medial side of the fifth metacarpal. This muscle takes partial origin also from the ulna. (5) The *supinator*, which has a deep layer arising from the ulna and which inserts on the radius by winding around the bone posteriorly, crossing the lateral surface to attach to the anterior surface. All of these muscles are supplied by the *radial nerve* or by its *deep branch*, the continuation of which, after passing through the supinator muscle, is known as the *posterior interosseous nerve*.

The medial epicondyle gives rise to many of the anterior forearm muscles that flex the wrist and fingers and pronate the hand. These are (1) the *pronator teres*, which has a deep head of origin from the ulna and which inserts on the lateral surface of the middle third of the radius; (2) the *flexor carpi radialis*, which inserts on the anterior (palmar) surfaces of the bases of the second and third metacarpals; (3) the *palmaris longus* (when present), which inserts on the flexor retinaculum and the apex of the palmar aponeurosis; (4) the *flexor carpi ulnaris*, which has a deep head from the ulna and inserts on the pisiform bone and indirectly by ligaments into the fifth metacarpal and (5) the *flexor digitorum superficialis*, a portion of which originates from the radius. This muscle sends tendons to the palmar surface of the middle phalanx of the medial four fingers. All of these muscles are supplied by the *median nerve*, except the flexor carpi ulnaris, which is supplied by the *ulnar nerve*. The median nerve passes between the two heads of the pronator teres, and the ulnar nerve between the two heads of the flexor carpi ulnaris.

The flexor and extensor carpi ulnaris muscles working together adduct the hand, whereas the flexor and extensor carpi radialis muscles function together to abduct the hand. The two flexors steady the wrist in extension of the fingers, i.e., they work as synergists, and the extensors steady it during flexion of the fingers. The student will recall that there is an extensor carpi radialis longus muscle as well as the brevis.

CASE 2–3: FRACTURE OF THE HUMERUS

History

An 11-year-old boy fell out of his tree house and fractured the left humerus about midway between the two ends of the bone. The left hand immediately assumed a flexed position and, regardless of effort, the boy could not extend the hand at the wrist or any of the fingers, including the thumb. In addition, flexion and supination of the left forearm were weakened but not abolished. Abduction of the thumb was greatly weakened, but flexion, adduction and opposition of this digit were unaffected. Extension of the forearm also was weakened somewhat. A neurologist observed that the boy had lost cutaneous sensation over part of the posterior arm, over most of the posterior forearm and over the dorsum of the left hand from the thumb to the middle finger. The boy was scheduled for neurosurgery, followed by orthopedic manipulation of the broken bone. The arm was put in a cast, which was removed 8 weeks later. During this time he was observed at regular intervals by the neurologist, who noticed that both the ability to extend the wrist and fingers and cutaneous sensation returned steadily. A year after the accident, the boy had regained complete control of the left upper limb and reported no sensory loss.

Questions

What structure other than the humerus was damaged and what is the term for the clinical condition thus produced?

Review the specific osteology of the humerus that is related to this injury. Discuss the pattern of musculocutaneous supply of the damaged structure with regard to the losses observed by the neurologist.

Does any other structure accompany the damaged one in its course around the humerus? If so, would you expect it to be damaged also?

Discuss briefly and in general terms the arterial blood supply to the axilla and the arm.

What are the alternate pathways by which arterial blood could travel to the site of the injury in the event that the main pathway was damaged?

Discussion

The humerus, in breaking, severed the *radial nerve,* which comes into contact with the bone in what is known as the *spiral groove.* The clinical condition that results is known as *wrist drop,* since the radial nerve supplies all of the extensors of the wrist and fingers (see below). Thus, the wrist and fingers go into an exaggerated flexed position due to the now unopposed flexors of these structures. The radial nerve, which is the direct continuation of the posterior cord of the brachial plexus, lies at its origin medial to the humerus (Fig. 2–3). It then winds posteriorly around the humerus from medial to

Fig. 2–3.—Relationship of the *radial nerve* and *profunda brachii artery* to the spiral groove of the humerus. The nerve is especially vulnerable to damage via humeral fractures where it comes into direct contact with the bone. (Redrawn from Tobin.[1])

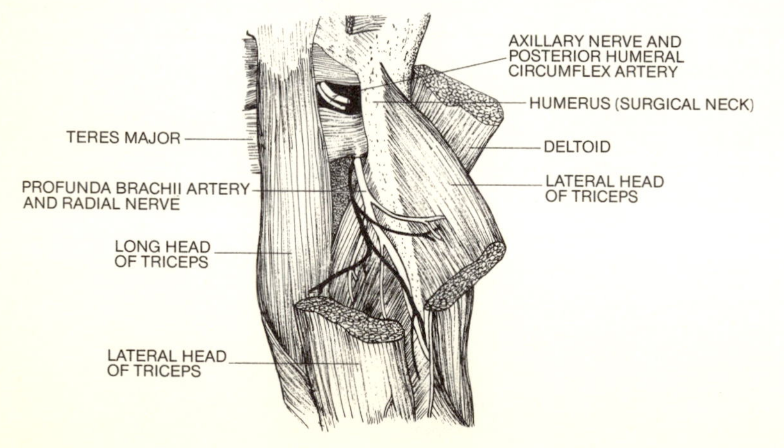

lateral. In this position it occupies the spiral groove, which, as its name implies, spirals around the humerus, occupying about the middle third of the bone. The nerve is particularly susceptible to damage where it lies in direct association with the bone. It pierces the posterior surface of the lateral intermuscular septum and emerges in the lateral aspect of the distal portion of the arm just above the superior boundary of the cubital fossa. Above this point the radial nerve gives off *muscular branches* to the *triceps* and the *posterior cutaneous nerve of the arm.*

Extension of the forearm was weakened, but not lost, because the branches of the radial nerve to the long and lateral heads of the triceps generally emerge from above the site of this lesion. Therefore, it was largely the function of the medial head that was lost. In the lower lateral portion of the arm, the main trunk of the nerve supplies branches to the *brachioradialis* and the *extensor carpi radialis longus* and gives off the *posterior cutaneous nerve of the forearm.* It also gives a branch to the anconeus. The brachioradialis muscle is actually a flexor of the forearm, and its loss explains the somewhat reduced strength of this action. Such a reduction would be barely noticeable, however, due to the relative strength of the biceps brachii and brachialis muscles, which are supplied by the musculocutaneous nerve. The main trunk of the nerve then divides into its two terminal branches, the *superficial* and *deep branches of the radial nerve.* The superficial branch, which is entirely sensory, courses under the medial border of the brachioradialis muscle to the wrist. It has no branches in the forearm. At the wrist it winds superficially over the tendons of the abductor pollicis longus and extensor pollicis brevis and longus muscles. This is one of the few places in the body where cutaneous nerves can be felt. Running a fingernail down one of these tendons will attest to this fact: if the pressure is hard enough, the nerves will discharge, as evidenced by a tingling sensation over the backs of the first two-and-one-half fingers, the area of cutaneous supply of the superficial branch.

The deep branch of the radial nerve supplies the *extensor carpi radialis brevis* and most of the other muscles that arise by a common tendon from the lateral epicondyle of the humerus (see Case 2–2). The actions of almost all of these muscles are described by their names so no additional discourse on this aspect is required here (the anconeus assists the triceps in extending the forearm). The deep branch thus supplies the *extensor digitorum,* the *extensor digiti minimi* and the *extensor carpi ulnaris.* It enters the *supinator muscle,* which it also supplies, occupying a position between the superficial and deep heads of this muscle. As it emerges from the supinator, its name is changed to the *posterior interosseous nerve.* As such, it accompanies the artery of the same name down the deep aspect of the posterior forearm to the back of the wrist where it terminates as articular branches. Before doing so, it supplies the *abductor pollicis longus,* the *extensor pollicis brevis,* the *extensor pollicis longus* and the *extensor indicis muscles.* Total loss of abduction of the thumb was avoided, since the abductor pollicis brevis muscle is supplied by the median nerve (see Case 2–1), as are the flexor pollicis brevis and opponens pollicis. The adductor pollicis is supplied by the deep branch of the ulnar nerve in the palm. Complete loss of supination was avoided, since the biceps brachii is also a supinator of the forearm and hand. All of these functions were eventually restored to normal by rejoining the two cut ends of the nerve. According to Marks,[2] "Suture of the divided radial nerve is more often successful than with any other divided upper limb nerves, in view of its predominant motor function."

The structure that accompanies the radial nerve in its course around the humerus in the spiral groove is the *profunda brachii artery,* the first branch of the brachial artery in the arm (see Fig. 2–3). There is little doubt that this structure was damaged also when the humerus was fractured. The arterial supply to the upper limb is derived from the subclavian artery. This large vessel becomes the *axillary artery* at the lateral border of the first rib and, in turn, becomes the *brachial artery* at the lower border of the *teres major muscle.* The axillary artery is subdivided into three parts by the pectoralis minor muscle, which crosses its middle third. Although variation is common, the following pattern usually is observed. The first part (medial to the muscle) has one branch, the *supreme* or *highest thoracic artery*. The second part (behind the muscle) has two branches, the *thoracoacromial trunk* and the *lateral thoracic artery*. The third part (lateral or distal to the muscle) has three branches, the *subscapular artery* and the *anterior* and *posterior circumflex humeral arteries*. The latter encircle the surgical neck of the humerus, anastomosing with each other and therefore providing potential collateral pathways to the surgical neck, which is a frequent site of breakage of the humerus. The profunda brachii artery sends an *ascending* or *deltoid branch* superiorly to anastomose with the humeral circumflex arteries, which constitute one collateral pathway for blood to reach the site of the injury in the present case (i.e., by flowing retrograde down this ascending branch). The profunda follows the radial nerve to the lateral surface of the arm where it divides into *radial* and *middle collateral arteries* that descend to occupy positions anterior and posterior, respectively, to the lateral epicondyle. The *radial recurrent artery* (a branch of the radial artery) anastomoses with the radial collateral, and the *interosseous recurrent artery* (a branch of either the common or posterior interosseous arteries) anastomoses with the middle collateral artery. These constitute a second collateral pathway for arterial blood to reach the injured area, again via retrograde flow. The anastomoses around the elbow (there are branches around the medial epicondyle also) ordinarily allow blood to flow around the brachial artery when it is "crimped" by protracted flexion of the elbow, but, in addition, they may play the important role of supplying injured areas, as in this case. They also may come into play in thrombosis of the brachial artery or in mechanical occlusion caused by tumors or other sources.

CASE 2–4: INFECTION OF THE HAND

History

A 49-year-old migrant farm worker cut his right thumb on an old piece of barbed wire. Having no available medical facilities, he simply wrapped his thumb in a strip of dirty bandana and went back to

I Googled Horace Wilbur. The site proclaiming his sex offender status was one of the first hits. How had we not known? How had we let him spend hours upon hours, in the quiet reading room, with his delicate detailing brushes, leaning over the shoulders of children like the one he had murdered?

The newspapers called even before the police. How many children had taken his classes? What ages? Had we even bothered to research him? The question I heard over and over, even though no one said it aloud, was, *How could you?*

The worst part was that they were right, these voices in my head. In all the time I saw myself as providing a sanctuary, I never thought of the boundaries that had to be maintained to keep that space safe. I never thought of the people who might want access to it for easy prey.

I began to prepare the speech I would give to my staff when I told them all that had happened. I sat down and began to compile a list of my volunteers' phone numbers.

I sat at my desk looking at a chain of emails from the writer who had recently moved to the area, looking for his phone number. He had told me to use "they" or "he" as his pronoun, but unable to bring myself to use "they" in the singular because of my respect for the English language, I had used the latter. He said he understood, and offered it as an option because he knew it was easier for people.

Quinn had moved to the area from San Francisco, somewhat bafflingly. He was an activist and a published author, with several books put out by small presses. We'd been delighted when he stopped by the library and suggested a program to promote literacy in young people through writing.

It was the opportunity I'd been wanting for some time. There was another artist in town, a musician, who had similar inclinations to creating social good. I chatted with Quinn on the phone several times about how perhaps the two of them could create a program that took in all of the lost sheep in the town. Our county has the highest illiteracy rate in work. The wound healed shortly thereafter, but within a few days his thumb was inflamed, swollen and exceedingly tender. It throbbed with pain almost continually unless held in an upright position. Fearing he would lose his job if he took time off to seek medical care, he continued to work and tried not to think of his thumb. With time, the pain and swelling in the distal portion of the digit subsided and was replaced by a more diffuse discomfort perceived along the area of the first metacarpal. This, in turn, seemed to proceed toward the man's wrist where it became localized and quite severe again for a few days. This was followed once again by a more diffuse but throbbing pain that worked its way up the forearm. Finally, a large area of inflammation appeared in the upper forearm in the area of the head of the radius. Within a few days, the infection eroded the skin as a large fulminating abscess, exuding great quantities of pus. With the release of the pus, the throbbing pain finally subsided. The man was taken to a local hospital where the wound was cleaned and treated. He was given an intramuscular injection of antibiotic, a prescription for oral administration of the drug and a topical antibiotic for the wound itself. He recovered quickly from the disorder, although from that time on movements of the right thumb were somewhat stiffer than normal.

Questions

In what specific structure of the thumb was the infection located that allowed it to travel proximally into the wrist?

If the infection had been in some other finger would the same course of events have been expected to occur?

Describe these structures as they relate to the long tendons in the palm of the hand.

Describe the arrangements of the deep fascia in the palm that might tend to inhibit the spread of infections in different circumstances.

What accounted for the stiffness of the man's thumb?

Discussion

The infection entered the *synovial sheath* of the man's right thumb, which is prolonged into the wrist approximately 2 cm proximal to the flexor retinaculum (Fig. 2–4). The long tendons from the flexor muscles of the forearm are enclosed in these synovial sheaths as they enter the palm deep to the flexor retinaculum. The sheath for the flexor pollicis longus mus-

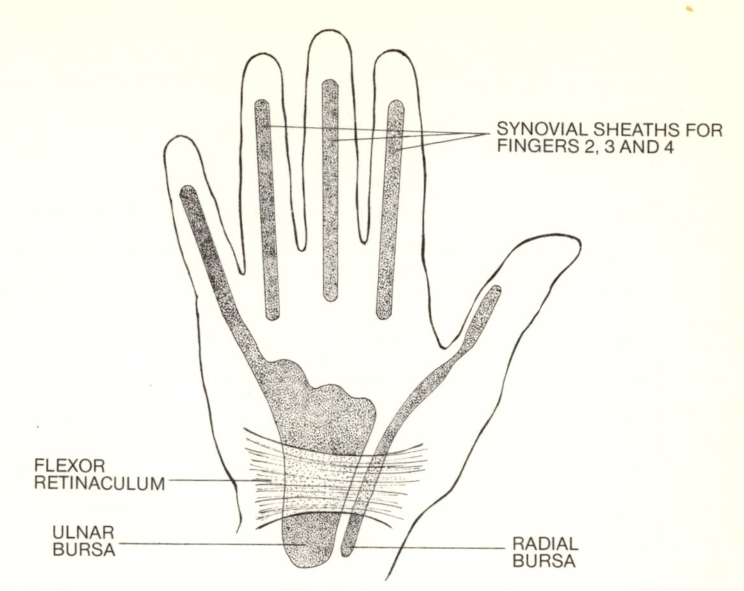

Fig. 2–4.—The *ulnar* and *radial bursae* of the hand and wrist. An infection in the thumb or little finger may spread into the palm and wrist via these bursae, whereas one in the second, third or fourth fingers is likely to be sequestered there. (Redrawn from Gardner *et al.*[3])

cle surrounds this tendon only and extends from the wrist to the tip of the thumb. It is frequently referred to as the *radial bursa,* and it is within this long tube that the infection in our patient spread from the tip of the thumb to the wrist. The tendons of the flexor digitorum superficialis and profundus muscles plus the median nerve are combined in a common synovial sheath, which also passes deep to the flexor retinaculum. It is known as the *ulnar bursa,* since it occupies that side of the wrist and forearm. An offshoot of this bursa extends to the tip of the little finger. The synovial sheaths for the second, third and fourth fingers do not connect with either of the bursae that extend into the wrist; rather, they begin near the tips of these digits and end blindly near the heads of the respective metacarpals. Thus, an infection in the little finger also could spread into the wrist via the synovial sheath for this digit, but one in the index, middle or ring fingers would tend to be sequestered in the respective synovial sheaths distal to the palm.

The infection in the radial bursa was restricted temporarily at the proximal end of that structure (this explains the localized severe pain in that area) but, as the pressure became great enough, the synovial membrane ruptured, allowing the infection to flow into the forearm. It was free to spread between the long flexor tendons and the pronator quadratus muscle up into the anterior forearm. Eventually it could spread no further (due to the tightly compact muscles of the proximal forearm), and so it broke through the subcutaneous tissue and the skin. Thus, a huge abscess near the elbow was the result of a small cut on the thumb, long since healed.

The synovial sheaths are mesothelial-lined membranes that have "visceral" and "parietal" layers much like the pleura and peritoneum (although they usually are not described in these terms). The visceral layer surrounds the tendon directly, whereas the parietal layer is separated from it by a potential space or cavity filled with synovial fluid. The two layers are connected at a reflection known as a *mesotendineum* (compare with the various peritoneal mesenteries). Blood vessels are transmitted to and from the tendons in specializations of the mesotendinea known as *vincula. Vincula brevia* are roughly triangularly shaped structures located near the points where the long tendons insert into the phalanges; *vincula longa* are long strands that can be identified by gently pulling the long tendons away from the bones of the digits. The synovial sheaths allow the tendons to glide with a minimum of friction (much like sliding two plates of glass on each other when they are separated by a film of liquid). In this case, the infection caused a scarring of the synovial membrane of the radial bursa that resulted in increased friction during movements of the flexor pollicis longus tendon and, consequently, stiffness of the thumb. The situation probably would have been much more debilitating had the infection been in the ulnar bursa, since this encloses the long flexor tendons of the remaining four fingers as well as the median nerve.

The deep fascia of the distal part of the anterior forearm and of the palm is specialized in many ways; it is important to note, however, that all the specialized structures are derived from the same basic layer of tissue. One of these is the *flexor retinaculum,* which bridges the concavity of the carpal bones and with them forms the *carpal tunnel* (see Case 2–5). Another is the *palmar aponeurosis,* which spreads out over the palm and gives offshoots that overlie each of the digits. The palmaris brevis muscle takes origin from the medial border of the palmar aponeurosis and runs medially to insert on the skin of the palm. Upon contraction (this can be reflexly demonstrated by pressing on the pisiform bone), it serves to deepen the concavity of the palm. The palmaris longus muscle, when present, inserts on the apex of the palmar aponeurosis. The deep fascia continues medially and laterally from the margins of the palmar aponeurosis over the hypothenar and thenar eminences, respectively. Between them, septa extend deeply to the first and fifth metacarpals (these have been termed the *lateral* and *medial palmar septa*). As a result, the palm is subdivided into *thenar* and *hypothenar spaces,* separated by the *central* or *intermediate palmar space.* An additional septum (the *intermediate palmar septum*) extends obliquely deep to the third metacarpal, thus dividing the central or intermediate area into the *medial* and *lateral deep palmar spaces.* These "spaces" contain fat and, therefore, are only potential in nature; they are retrotendinous, that is, they lie deep to the long flexor tendons. Their significance lies in the fact that infections that may accumulate in these spaces tend to be localized there. In other words, due to the nature of the septa, infections usually do not spread from these spaces, and they are receptive to treatment merely by incising and draining the pus. Severe infections may break out of these spaces, however, and enter the forearm between the long flexor tendons and the pronator quadratus muscle. It also is important to note that infections may exceed the confines of the synovial sheaths and enter these deep spaces within the palm.

Final specializations of the deep fascia of the palm are collectively the *fibrous flexor sheaths.* These lie over the long flexor tendons and their synovial sheaths as they cross the phalanges en route to their insertions. The flexor tendons are retained in place by these sheaths, which prevent "bowstringing" of the tendons upon flexion of the fingers. Each sheath is strengthened by a fibrous band opposite the center of the proximal and middle phalanges but is thin opposite the interphalangeal joints.

It also should be emphasized that the skin of the palmar surface of the hand is bound down to deeper tissues by connective tissue septa so that various "pads" are formed (separated by the creases or "lines" of the palm), which assist in grasping.

CASE 2–5: LOSS OF NORMAL HAND FUNCTION

History

A 55-year-old dentist began to experience certain difficulties with the manipulation of his right hand. Specifically, he noticed a progressive lack of coordination and strength in the thumb and also a reduction in the ability to perform fine movements of the index and middle fingers. He observed as well a tingling sensation (paresthesia) and reduced tactile sensation (hypoesthesia) in the first two-and-one-half fingers (primarily on the palmar surfaces). As the condition progressed, he began to perceive pain from the palmar aspect of the lateral half of the hand. Within a few months of the onset of symptoms, it had become obvious to him that he could no longer successfully practice dentistry under these circumstances. He consulted an orthopedic surgeon, who, after a thorough examination, concluded that there was fibrosis of the flexor retinaculum of the patient's right wrist and that surgery was the only treatment that could provide relief from the condition. This was performed, and within several weeks the dentist was back at work with no detectable loss of manual dexterity.

Questions

What is the name of this condition?

Compression of what structure was responsible for the observed symptoms? Describe its distribution.

What other structures occupy the same general area as they enter the hand?

Identify the carpal bones with regard to their position and any outstanding characteristics they may demonstrate.

Describe the flexor retinaculum and its connections in the wrist.

What is the function of the lumbrical muscles?

Discussion

The patient in this case was suffering from *carpal tunnel syndrome*, the result of compression of the *median nerve* in the carpal tunnel. The latter is a passageway between the *flexor retinaculum* and the concave surface presented by the palmar aspects of the *carpal bones*. The other contents of the carpal tunnel are the four tendons of the flexor digitorum superficialis muscle and the four tendons of the flexor digitorum profundus. These structures are enclosed in synovial sheaths, collectively termed the *ulnar bursa* (see Case 2–4). The tendon of the flexor pollicis longus muscle also traverses the carpal tunnel in its own synovial sheath, the *radial bursa*. Occasionally, fibrosis of the retinaculum reduces the volume of the carpal tunnel, resulting in compression of the median nerve and in the observed symptoms. This condition also can be caused by a tumor or other mechanical obstruction of the tunnel.

The median nerve, as it enters the palm of the hand just beyond the distal border of the flexor retinaculum, divides into its two terminal branches, the *lateral* and *medial divisions* (see Fig. 2–1). The lateral division gives off the *recurrent (superficial) branch* to most of the thenar muscles and then splits into the *palmar digital branches* to the two sides of the thumb and the lateral surface of the index finger. The latter are sensory in nature. The distribution and function of the recurrent branch were described in detail in Case 2–1; its loss (or interference due to compression, as occurred in this case) explains the loss of coordination and strength of the dentist's right thumb. The lateral division also supplies motor fibers to the first lumbrical muscle. Although variation in the nerve supply of the hand is not uncommon, the description given here is probably the typical one. The medial division sends palmar digital nerves to the medial side of the index finger, to both sides of the middle finger and to the lateral half of the ring finger. The remaining one-and-one-half fingers on the palmar aspect are supplied by the ulnar nerve. The median nerve also variously supplies the dorsal surfaces of the distal digits of the lateral fingers (i.e., the fingernails and perhaps some skin). It is important to note that there is considerable overlap in the areas of cutaneous supply of these nerves as well as most other sensory nerves in the body. Thus, denervation of a single nerve often results in hypoesthesia rather than anesthesia of a given area of skin. The medial division of the median nerve also gives a twig to the second lumbrical muscle. We can now see, on the basis of the distribution of this nerve, the explanation for the dentist's symptoms.

It is not the purpose of this book to encourage the student to memorize all of the aspects of each of the eight carpal bones. Nevertheless, the student should have an understanding of their relative positions within the wrist and be able to identify them in their normal arrangement from both the palmar and dorsal aspects. The carpal bones are arranged in two rows, proximal and distal. From lateral to medial, the proximal row consists of the *scaphoid, lunate, triquetrum* and *pisiform* bones. The pisiform actually occupies a position on the anterior surface of the triquetrum. In the distal rows the order from lateral to medial is the *trapezium, trapezoid, capitate* and *hamate.* The scaphoid, which presents a palpable tubercle, is the largest bone in the proximal row and is the most susceptible to fracture of all of the carpal bones. Nonunion after fracture of the body of this bone is characterized by pain and swelling in the "anatomical snuffbox" (the area between the extensor pollicis brevis and extensor pollicis longus tendons). The scaphoid forms the floor of the "snuffbox." The capitate is the largest of all of the carpal bones. The hamate is probably the easiest of the carpal bones to identify due to its readily recognizable "hook" on the palmar surface. The carpus is concave anteriorly.

The flexor retinaculum is a quadrangular specialization (about 3 square cm in area) of the deep fascia of the anterior forearm that bridges over the carpal concavity and thus converts it into a tunnel. Proximally, the retinaculum is attached to the tubercle of the scaphoid laterally and the triquetrum and pisiform medially. Distally, it is anchored to the trapezium laterally and the hook of the hamate medially. In the dentist it was the flexor retinaculum that became fibrosed and thus compressed the carpal tunnel; simple surgical incision of this structure relieved the pressure on the median nerve and thus abolished the neuromuscular disorder.

The *lumbrical muscles* originate from the lateral sides of the four deep flexor tendons in the hand and insert on the extensor expansion or hood of the medial four fingers. The first and second lumbrical muscles (to the index and middle fingers, respectively) are supplied by the median nerve, as we have seen. The third and fourth lumbricals (to the ring and little fingers, respectively) are innervated by the deep branch of the ulnar nerve. Since the ulnar nerve crosses the flexor retinaculum superficially to enter the palm, it is not injured in the carpal tunnel syndrome. The lumbricals have a unique function in that they flex the metacarpophalangeal joint (owing to their course anterior to this joint axis), whereas they extend the interphalangeal joints (owing to their insertion on the extensor expansion). They are therefore capable of producing the so-called "z" position of the hand (flexed metacarpophalangeal joints but straight fingers) and are important in such motions as the upstroke in writing. Some hand surgeons feel that an understanding of the actions of these small muscles is a key to understanding the over-all function of the hand.

CASE 2–6: KNIFE WOUND OF THE FOREARM

History

A young man appeared at the emergency room of a large city hospital following a gang fight in a downtown ghetto. He had suffered a deep knife wound at the anteromedial aspect of the right forearm, approximately 3 inches proximal to the wrist. The bleeding (which had been only moderate) was stopped, the wound was cleaned and stitched, and the young man was released from the hospital with instructions to return in 1 week for examination and to have the stitches removed. He neglected this advice, however, and when the skin around the wound became inflamed and began to itch, he removed the sutures with his own knife. The inflammation subsided and he went about his business, but he soon recognized that his "right hand didn't work anymore as well as it used to." He noticed that his hand had assumed somewhat the look of a claw, with exaggerated extension of the metacarpophalangeal joints and flexion of the interphalangeal joints, primarily of the medial two fingers. The fingers could not be abducted or adducted, and adduction of the thumb was lost. When he attempted to pick up a piece of paper between thumb and index finger, the thumb would assume a hyperflexed position at the interphalangeal joint; this is known as Froment's sign or Froment's paper sign, and it is characteristic of this disorder. In addition to these problems, the patient noticed sensory loss over the medial aspect of the palm and dorsum of the hand and on the front of the ring and little fingers and on the back of the medial two-and-one-half fingers. He did not seek additional medical attention. The disorder was permanent, and was responsible for a physical deferment from military service as well as causing him great difficulty in handling manual labor jobs thereafter.

Questions

What structure was injured and what is the name of the condition thus produced?

Trace the course of this structure from its origin into the hand, noting its basic positions and relationships.

Discuss the reasons for the specifically noted symptoms of the disorder from a mechanical point of view.

Would the symptoms have been different had the structure been injured at a different point in the upper limb, e.g., just proximal to the elbow?

Could the patient's condition have been significantly improved had he sought further medical treatment at an early point in time?

Discussion

The structure that was injured was the *ulnar nerve* and the condition it produced (as was physically evident even to the patient) is known as *clawhand* or *main en griffe* (Fig. 2–5, A). In some individuals the symptoms of the medial two fingers are exaggerated and, when the index and middle fingers are normally extended, the result has been referred to as the "*hand of benediction*" (Fig. 2–5, B).

The ulnar nerve (C_7; T_1) is one of the terminal branches of the medial cord of the brachial plexus. It descends in the arm (in which it has no branches) medial initially to the axillary artery and later to its continuation, the brachial artery. At about the middle of the arm it passes from the anterior to the posterior osteofascial compartment by piercing the medial intermuscular septum. It descends over the medial head of the triceps and then occupies the groove on the back of the medial

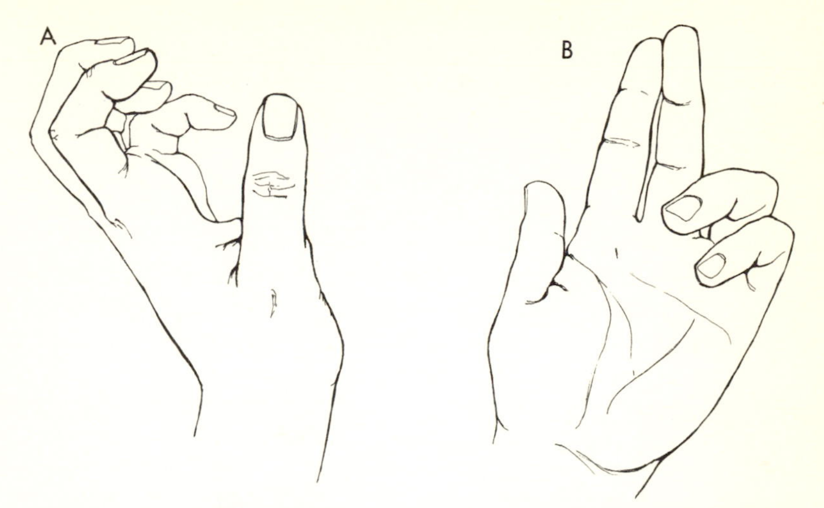

Fig. 2–5.—**A,** clawhand. **B,** "hand of benediction." (Based on Wilson, F. C.: *The Musculoskeletal System* [Philadelphia: J. B. Lippincott Co., 1975].)

epicondyle of the humerus; at this point it is superficial in position, being susceptible to blows (we speak of hitting the "funny" or "crazy" bone) or to injury. After giving a twig to the elbow joint, it enters the forearm by passing between the two heads of the *flexor carpi ulnaris muscle,* both of which it supplies. It then comes to lie on the medial half of the *flexor digitorum profundus muscle,* this portion of which it also supplies. It descends through the forearm on this muscle and, at a point about 2 inches above the wrist, gives off the *dorsal (cutaneous) branch,* which supplies the medial aspect of the dorsum of the hand and the medial two-and-one-half fingers with sensory innervation. In the injured young man the main trunk of the nerve was cut above the origin of the dorsal branch, as evidenced by the loss of sensation in these areas. The *ulnar artery* accompanies the nerve in this part of its course but was apparently not cut, as demonstrated by the lack of severe bleeding at the injury site. The nerve then enters the palm by crossing superficial to the flexor retinaculum. At about the distal border of this structure it divides into *superficial* and *deep branches.* The superficial branch gives a twig to the *pal-*

partially burned, in a field near where the friend had lived. There had been a manhunt, and details had trickled out. He was found with all the necessary items to create a new life somewhere far away: a fake ID, boxes of hair dye, glasses he did not need.

The email that pinged my phone was about that, but also about more. It was from the Regional Director, saying that the man who killed the girl had been caught, and that she recognized him as a former painting instructor and volunteer at the library.

I sat down. The chairs in the reading room were hard. No one ever used them. No one except the old people, and the writer who had recently moved to town and volunteered to teach a children's class on short story composition.

Horace Wilbur didn't look like someone who would attempt to have sex with a thirteen-year-old girl, then shoot her at point-blank range when she refused. He was an older man in his fifties, with short, balding, greying hair and an affable smile. He taught watercolor to the teens after school. His paintings were of Ohio birds he'd seen in his backyard in a nearby, even more rural town. The Northern Bobwhite, the Yellow-Billed Cuckoo, the Mourning Dove—each feather painted in loving detail. *How could a man who painted doves shoot a teenager in the face?* I wondered, back in my office. I had met Wilbur. We chatted about the impact art could have on young lives, the incarcerated, the illiterate, those who were cast out by society. I had always wanted to help those people, and behind my desk, I told myself that I did. I had looked at people like Horace and seen someone in the trenches, someone spending their free time, unpaid, building a better world.

I called the Regional Director.

"Horace?" I said. My voice was shaking.

"Horace," she replied. "We're scrambling. "We're trying to put new policies in place before this comes back to us. Expect to have to go through a background check soon, along with all of your employees and volunteers. We should have seen this. We should have known."

She hung up before I could say much. I sat at my desk. The bright screen of the computer seemed to grow and shrink, but I knew that was just my own jagged breaths.

maris brevis muscle and then supplies the medial aspect of the palm and medial one-and-one-half fingers with sensory branches. The deep branch descends through the *hypothenar muscles,* which it supplies, deeply into the palm to accompany the deep palmar arterial arch. In this position it supplies all of the *interosseous muscles,* the third and fourth *lumbricals,* the *adductor pollicis* and the deep head of the *flexor pollicis brevis muscle.*

The *dorsal interosseous muscles* abduct the fingers (away from the plane of the middle finger), whereas the *palmar interossei* adduct them. A useful mnemonic device for the functions of these muscles is *dab* for *d*orsal *ab*duct, and *pad* for *p*almar *ad*duct; this is the only way that two sensible words can be constructed from these alphabetical components. Denervation of these muscles explains the loss of these actions. The intrinsic muscles of the little finger (hypothenar) are the *abductor digiti minimi,* the *flexor digiti minimi* and the *opponens digiti minimi;* their names imply their actions, which also were lost in the young man's case. Similarly, adduction of the thumb was lost, and flexion of the thumb was weakened. The latter reduction was barely noticeable, since the deep head of the flexor pollicis brevis usually is much weaker than the superficial head (which is supplied by the median nerve) and, of course, the flexor pollicis longus muscle was still intact. The denervation of the medial two lumbrical muscles is the key to explaining the clawing of the hand. These muscles originate from the long tendons of the flexor digitorum profundus muscle in the hand, cross the metacarpophalangeal joint anteriorly and insert on the extensor expansion or hood (see Case 2–5). Thus, they act to flex the metacarpophalangeal joint while at the same time, due to exerting tension on the extensor expansion, extending the interphalangeal joints. Denervation of these muscles explains the exaggerated extension of the metacarpophalangeal joint (due to the reduced opposition to the long extensors of the fingers) and the flexion of the interphalangeal joints (due to the reduced tension on the extensor expansion). The result is clawhand.

Clawing of the medial two fingers does not occur, or is much less pronounced, if the ulnar nerve is sectioned around the elbow, because the medial half of the flexor digitorum profundus, which produces interphalangeal joint flexion, is also denervated. In this case, adduction and flexion of the wrist also would be weakened due to denervation of the flexor carpi ulnaris muscle. A lesion of both the ulnar and median nerves in the distal forearm produces the so-called *complete* clawhand. In this case all of the metacarpophalangeal joints would be extended and all of the interphalangeal joints flexed since the first two lumbricals are supplied by the median nerve. It should be mentioned, however, that there still exists a degree of controversy regarding the relationship of function of the lumbrical muscles to that of the interossei and to the long flexors and extensors, and that the observed clawing in ulnar nerve lesions probably should not be attributed solely to denervation of the lumbrical muscles.

Had the ulnar nerve lesion been recognized during the young man's visit to the emergency room, the two cut ends of the nerve might have been surgically reconnected, resulting in some degree of regeneration and recovery of hand function. However, according to most sources, recovery after section of the ulnar nerve is rarely complete. Certainly, chances for any effective recovery of function were reduced to near zero when the young man neglected to return for medical care following emergency treatment.

CASE 2–7: ABNORMAL HIP FUNCTION

History

As the result of a suspected birth injury, a male child experienced difficulty in learning how to walk, lagging far behind the normal time limits for acquiring this trait. As he developed into a young boy, his walk was characterized by a lurching or waddling gait in which he would bend his trunk to the right in order to raise the left foot off the ground. When the left foot was planted, this unusual trunk action was not necessary in order to lift the right foot in the swing phase of locomotion. Clinical testing revealed no sensory loss but, rather, motor losses in the gluteus medius, gluteus minimus and tensor fasciae latae muscles. The condition was permanent.

Questions

What single structure was damaged in this case?

Trace the course and distribution of this structure.

Do you know the clinical term for the waddling or lurching gait characteristic of this injury?

Which side was injured, right or left?

Discuss the origins, insertions and actions of the three muscles involved and describe as well as you can from an engineering standpoint the relationship of the respective losses to the patient's disability.

Could he walk at all without lurching? Are there any mechanical or support devices that would assist the patient?

Discussion

The single structure that was damaged in this patient (probably during birth, which would explain his difficulties from infancy) was the *right superior gluteal nerve*. This nerve supplies motor innervation to the *gluteus medius*, *gluteus minimus* and *tensor fasciae latae muscles*. The gluteus maximus is supplied by the inferior gluteal nerve. The lurching or waddling motion characteristic of this disorder is known as the *Trendelenburg gait*, and the test for it (the observations and explanations for which will be discussed below) is referred to clinically as the *Trendelenburg test*.

The external or *gluteal surface* of the *ilium* presents three lines or ridges that are developed to a greater or lesser extent depending on the muscularity of the individual. Two of these lines (the anterior and inferior) cross the ilium transversely (horizontally), whereas the posterior one runs in a vertical direction. The *anterior* (sometimes called the middle) *gluteal line* arches between the tubercle of the crest of the ilium and the greater sciatic notch. The *inferior gluteal line* extends also in an arching course from the anterior inferior iliac spine to the greater sciatic notch below the anterior line. The *posterior gluteal* line extends from a point on the iliac crest about 5 cm anterior to the posterior superior iliac spine to the greater sciatic notch near the point where the anterior line joins it.

Any one or all of these lines may be difficult to identify on the skeleton.

Before describing the gluteal muscles, a word about the *gluteal fascia* is necessary in order to fully understand its attachments. The gluteal fascia surrounds the gluteus maximus, the most external and largest of the gluteal muscles, and then lies over the gluteus medius. It splits to enclose the tensor fasciae latae muscle. Due to its strength in various places, it frequently is referred to as the *gluteal aponeurosis,* especially where it covers the gluteus medius. Distal to the tensor fasciae latae muscle it extends as a thickened portion of the fascia lata known as the *iliotibial tract* (see below). Superiorly, this fascia is attached to the iliac crest and to the sacrotuberous ligament behind.

The *gluteus medius muscle* arises from the external surface of the ilium between the anterior and posterior gluteal lines and from the deep surface of the gluteal aponeurosis that overlies it. It inserts via a strong stout tendon on the lateral surface of the greater trochanter of the femur. The *gluteus minimus* arises from the area of the ilium between the anterior and inferior gluteal lines. It inserts on the anterior border of the greater trochanter. The *tensor fasciae latae* originates from the anterior superior iliac spine and from the adjacent area of the iliac crest. It is inserted into the iliotibial tract by virtue of the fact that the latter is formed at its distal extent by the conjunction of the two layers of gluteal fascia (aponeurosis) that surround the muscle. The iliotibial tract in turn is inserted into the lateral lip of the linea aspera via the lateral intermuscular septum and into the lateral retinaculum of the patella. As mentioned above, all three muscles are supplied by the *superior gluteal nerve* ($L_{4\text{-}5}$; S_1), which is a branch of the sacral plexus. The nerve, along with the superior gluteal artery, leaves the pelvis through the greater sciatic foramen above the piriformis muscle. These are the only structures that exit from the pelvis above that muscle; the majority of the remaining structures (e.g., sciatic nerve, inferior gluteal nerve and artery, pudendal nerve, internal pudendal artery) leave the pelvis through the greater sciatic notch below the piriformis. The superior gluteal nerve also supplies the hip joint.

The gluteus medius and minimus muscles abduct the thigh and rotate it medially (internally). When the thigh is fixed, they exert their force on the ipsilateral side of the pelvis, holding it level or even effecting an upward tilt when the contralateral foot is raised. In other words, rather than acting on the thigh, they exert their pull on the ilium in this case (Fig. 2–6). If this action were not possible, for example, if these muscles were paralyzed, as in our patient, the contralateral portion of the pelvis would sag upon lifting that foot, and the individual

Fig. 2–6.—Biomechanics of the hip joint. The *abductor muscles* keep the contralateral side of the pelvis from sagging when the contralateral foot is raised off the ground. If these muscles are paralyzed, the result is the Trendelenburg (lurching) gait. (Based on Wilson.)

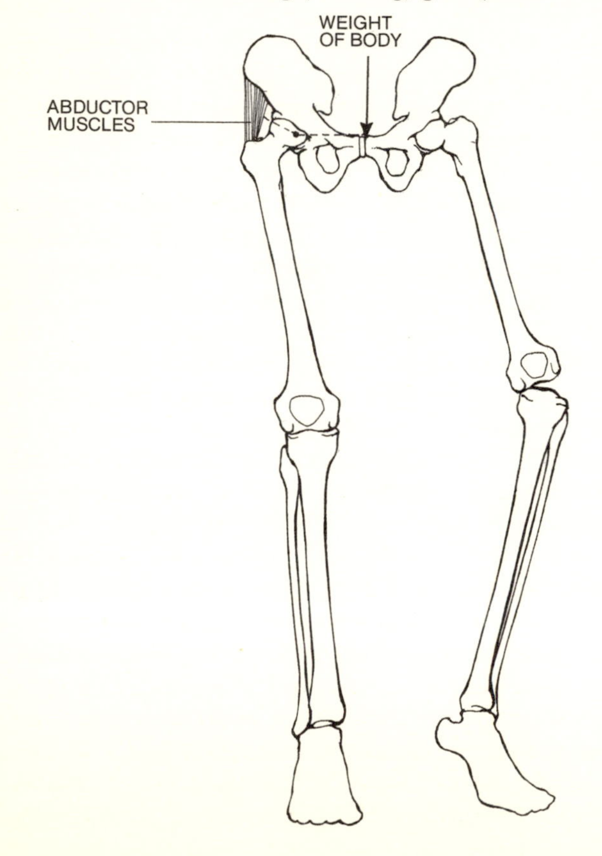

would probably fall over to that side. This observation explains, therefore, the necessity of the trunk to lurch laterally to the side of the paralysis in order to tilt the pelvis enough to allow the contralateral foot to clear the ground in ordinary walking. The result is the characteristic waddling Trendelenburg gait. This also has been referred to as an *abductor lurch,* the reasons for which are now clear.

A concise and simple biomechanical explanation of this condition has been given by Rosse and Clawson.[4] They note that the *center of mass* of the body occurs at about the middle of the second segment of the sacrum. The distance from the hip joint to the center of mass is approximately three times the distance from the hip to the point of origin of the gluteus medius and minimus muscles. Thus, the abductor force necessary to keep the pelvis level when one foot is raised is three times the body weight. During ambulation the compression force on the hip joint is about four times the weight of the body. When the abductors on one side are weak or paralyzed, the center of mass must be moved over toward the paralyzed side (lurch) in order to eliminate the abductor force necessary to balance the pelvis. The authors state, "The hip compression force then becomes equal to the body weight alone. Therefore, trunk lurches to the side of the stance leg are compensatory for abductor muscle weakness or for hip pain whatever the cause." They note that the use of a cane in the hand opposite the weak abductors compensates for this weakness or for hip pain and eliminates the trunk lurch. It should also be pointed out that the patient, in the absence of some artificial support, will often allow the contralateral hemipelvis to sag during a resting stance.

The gluteus maximus as well as the tensor fasciae latae insert on the iliotibial tract. The two muscles, the iliotibial tract and the lateral intermuscular septum, combine to form "a continuous, strong, musculoligamentous apparatus which is an important mechanism in maintaining posture and in locomotion."[3] This can be tested by placing the fingertips on the lateral aspects of the thighs while simply assuming a standing posture. With minor changes in attitude (e.g., slight swaying), the iliotibial tracts will be felt to alternately tense and relax. In a sense, they act much like guy wires in maintaining the upright posture of the body, especially in resistance to laterally applied forces. Denervation of the tensor fasciae latae muscle, as occurred in this patient, would create some, but not total, disability in this regard since the gluteus maximus muscle was still intact.

CASE 2–8: BACKACHE AND LOWER LIMB PAIN

History

A 44-year-old stevedore observed the gradual onset of recurrent low backache characterized by intervals in which he experienced complete freedom from pain. This went on for a period of several months until, upon lifting a particularly heavy load, the back pain became severe and was accompanied by pain extending down the posterolateral surface of the right thigh and leg and onto the dorsum of the right foot. His discomfort increased with activity but became mild to absent upon lying down. Physical examination revealed definite restricted forward flexion of the trunk and tenderness in the upper portion of the buttocks and along the lower portions of the erector spinae and deeper paraspinal muscles. In the supine position the right lower limb with knee extended ("straight leg") could only be raised about 45 degrees from horizontal before causing increased pain in the back and limb (in normals, the straight leg may be raised to 90 degrees without causing pain and, in individuals with backache only, this figure is reduced by only 20–30 degrees). Inspection revealed slight wasting of the right calf muscles. There was weakness of plantar flexion at the ankle and toes and of inversion of the foot. There was no apparent loss of tendon reflexes (knee jerk and ankle jerk), but some sensory loss was detected in a narrow strip of skin over the posterolateral area of the right lower limb. Radiologic examination revealed a narrowing of the normal dimensions of the space between the fourth and fifth lumbar vertebrae (L. V. 4 and L. V. 5). The patient was prescribed a regimen of complete bed rest for a period of 3 weeks combined with analgesics. He was not allowed to perform any heavy lifting for a period of 6 weeks after being allowed out of bed. After this time he was permitted to return to work and had not experienced any additional difficulty more than 3 years after the initial incident.

Questions

On the basis of the information given, pinpoint the site and the nature of the patient's problem.

Describe a typical vertebra and discuss its articulations with adjacent vertebrae and its relationships to other structures, mainly spinal nerves.

Describe the basic differences between cervical, thoracic and lumbar vertebrae.

What is the basic pathology of this condition? When is it most likely to occur?

Discuss the relationship of the disorder to the neurologic signs observed and describe the rationale for the various physical symptoms demonstrated by the patient.

What is the rationale for the conservative type of treatment prescribed?

Discussion

The patient's condition was caused by prolapse of the *intervertebral disk* between L. V. 4 and L. V. 5 (in lay terminology this is a "slipped" disk). The details of this disorder will be described after careful consideration of the normal anatomy of vertebrae and their relationships with each other, with the spinal nerves that exit between them and with the intervertebral disks that occupy the spaces between them.

The *vertebral column* extends from its articulation with the occipital condyles of the skull to the tip of the coccyx. It consists of seven *cervical*, 12 *thoracic* and five *lumbar* vertebrae, plus the *sacrum* and *coccyx*. The vertebral column provides support for the body and protection for the spinal cord. The latter terminates in the adult at the level of the intervertebral disk between L. V. 1 and L. V. 2. Below this level the spinal nerves trail inferiorly and exit between individual vertebrae in the lower lumbar, sacral and coccygeal areas. This pattern of nerves has been compared to a horse's tail and therefore has been called the *cauda equina*.

The so-called "typical" vertebra consists of a *body* and a *vertebral arch* (Fig. 2–7, A). The bodies of adjacent vertebrae are separated by the intervertebral disks that form one aspect of their articulation with each other. The vertebral arches are made up of *pedicles, laminae* and *spinous processes*. The pedicles and laminae surround the *vertebral canal* that houses the spinal cord. (An operation in which the laminae are cut bilaterally and the segment of the arch removed in order to expose the spinal cord is known as a *laminectomy.*) *Transverse processes* extend laterally from the points of union of the pedicles and laminae. These form important connections for muscles that operate on the vertebral column (the intrinsic and extrinsic muscles of the back) or take their origins there and operate on other parts of the body, e.g., on the scapula (levator scapulae muscle) or the ribs (scalenus muscles). In the thorax the transverse processes also form one of the articular surfaces for the ribs. The vertebral arches also present *superior* and *inferior articular processes* on each side, with appropriate *superior* and *inferior articular facets* for articulation with adjacent vertebrae. A *vertebral notch* is present on the superior and inferior surfaces of each pedicle. Together, the two form an *intervertebral foramen* on each side for the transmission of spinal nerves from the spinal cord (Fig. 2–7, B.).

The first and second cervical vertebrae (C. V. 1 and C. V. 2) are known, respectively, as the *atlas* and the *axis,* and they present specific characteristics that differ from typical vertebrae (see a standard text). Characteristics that help to delineate cervical vertebrae from others include the presence of bilateral *foramina transversaria* in the transverse processes for transmission of the vertebral arteries to the cranial cavity. C. V. 7 sometimes does not possess these foramina and, if it does, the vertebral arteries almost never traverse them. The bodies of the cervical vertebrae are small but broad and the spinous processes are short and usually present bifid ends; the latter are often palpable. C. V. 7 is known as the *vertebra prominens* because of its prominent spinous process, which forms a noticeable "bump" on the back of the neck. The first pair of cervical spinal nerves exit the vertebral canal *above* the atlas; the remaining spinal nerves exit *below* the correspondingly numbered vertebrae, i.e., C_7 exits below C. V. 7.

Thoracic vertebrae may be distinguished from the others of the vertebral column primarily by the presence of *costal facets* for articulation with ribs. The facets for the *heads of the ribs* are located at the points where the vertebral arch joins the body. The facets for the *tubercles of the ribs* are located on the anterior surfaces of the ends of the transverse processes.

Lumbar vertebrae may be distinguished primarily by their relatively massive size (since they bear more weight than their cranially placed counterparts) and by the absence of foramina transversaria and costal facets. L. V. 5 is the largest in the body.

The *sacrum* is formed by the fusion of five vertebrae. Its superior articular surface is much the same as the inferior surface of the body of L. V. 5, with which it articulates via the corresponding intervertebral disk. It also presents *superior articular facets* for articulation with the inferior facets of L. V. 5. The *coccyx* is formed from three to five vertebrae. The upper one often is fused to the inferior aspect of the sacrum, and any of the others may be fused with one another. This may become important in childbirth, i.e., the coccyx sometimes needs to be artificially "broken" in order to increase the dimensions of the birth canal.

The bodies of adjacent vertebrae are held together by the *anterior* and *posterior longitudinal ligaments* (Fig. 2–7, C). It should be noted that the posterior longitudinal ligament lies within the vertebral canal, occupying a position just anterior to the spinal cord (anterior to the cauda equina below L_2). The intervertebral disks consist of a peripheral *anulus fibrosus* and a central *nucleus pulposus*. The former is composed of numerous lamellae of collagenous tissue arranged in a spiral fashion, with different lamellae demonstrating different fiber directions with respect to each other. The nucleus pulposus consists basically of collagenous fibers and a semigelatinous material that acts like a fluid. It is ordinarily kept in place by the anulus fibrosus and by cartilaginous plates that cover the superior and inferior surfaces of the disk. Although the two components of intervertebral disks subserve many functions, the primary one is to work together to create a shock-absorbing mechanism between adjacent bodies of vertebrae.

The intervertebral disks are subject to pathologic changes, especially in the lower lumbar area of the back where the loading strains are the greatest. If this loading force is too great, prolapse of the disk may occur, in which case the nucleus pulposus herniates partially or completely through the anulus fibrosus. This herniation may occur anywhere around the

HORACE WILBUR'S PAINTBRUSHES

Alex DiFrancesco

I WAS TIDYING UP THE QUIET READING ROOM WHEN MY PHONE vibrated with an incoming email alert. No one ever uses the quiet reading room, it's almost always empty. We mostly have teens from the middle school across the street who are anything but quiet. They take movies out, access our live stream services, and use our computers. The quiet reading room is used by the old folks who can't afford to have the morning paper delivered. We try to welcome all people, and all of their interests. I have always thought of the library as a safe haven for the kids who don't fit in on the football fields, the old pensioners, the quiet intellectuals at odds with this small, working-class town—for anyone, really, who needs it.

Of course, I had heard the news prior to that morning. A thirteen-year-old girl, missing, last seen with a family friend who turned out to be a registered sex offender. No one had known if the family knew about his past, or what the whole story was. Three days after the missing persons report had been filed, her body was found,

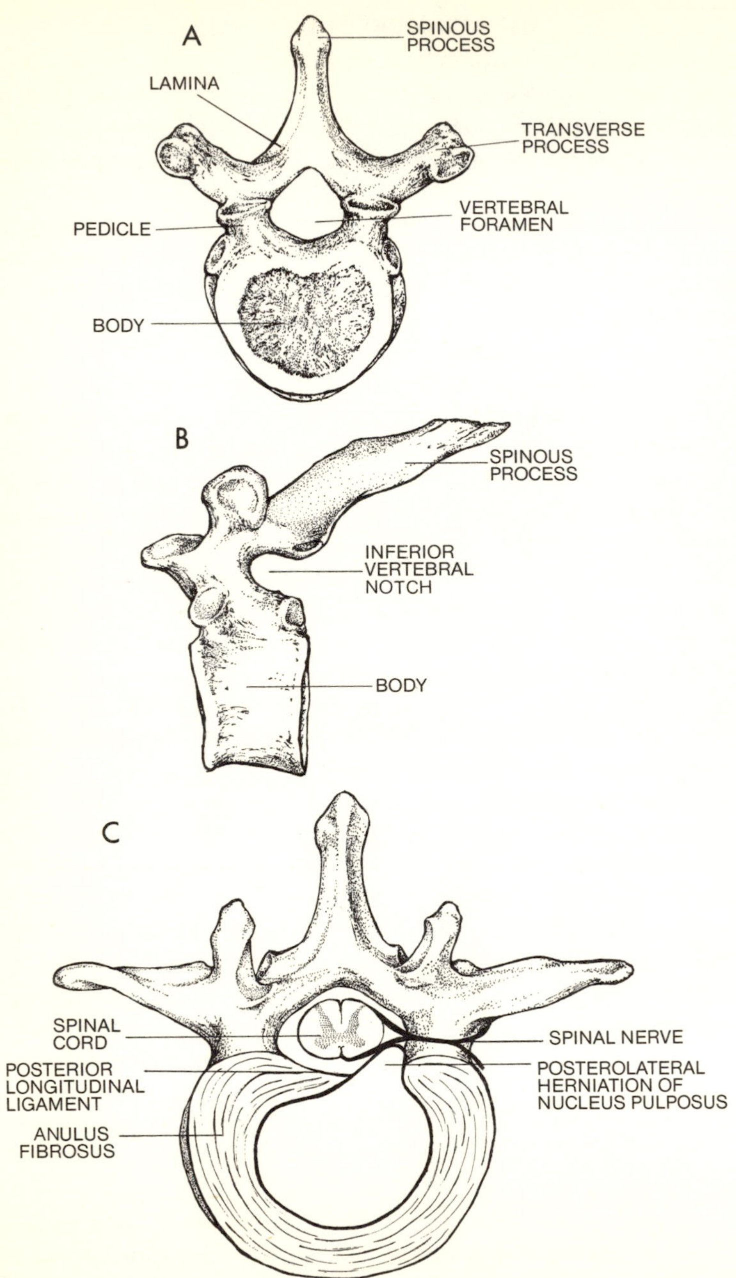
A
SPINOUS PROCESS
LAMINA
TRANSVERSE PROCESS
PEDICLE
VERTEBRAL FORAMEN
BODY
B
SPINOUS PROCESS
INFERIOR VERTEBRAL NOTCH
BODY
C
SPINAL CORD
SPINAL NERVE
POSTERIOR LONGITUDINAL LIGAMENT
POSTEROLATERAL HERNIATION OF NUCLEUS PULPOSUS
ANULUS FIBROSUS

periphery of the anulus but is more common in the areas where the longitudinal ligaments are deficient. Some investigators feel that the anulus is weakest posterolaterally. Lesser forces also may cause herniation if the disk has undergone degenerative changes. For example, a normal component of aging is a decrease in water content of the nucleus pulposus. In addition, in disk pathology, there appears to be an abnormally rapid breakdown in polysaccharide-protein linkages, with a reduction of the polysaccharide content of the nucleus pulposus plus an increased deposition of collagenous fibers.[5] As a result, there are changes in the gel structure of the nucleus that further alter its hydrostatic properties. This, combined with the formation of cracks or fissures in the anulus, frequently predisposes to disk herniation. Intervertebral disk prolapse is most frequent in individuals between the ages of 30 and 50. The fact that older people are less prone to lifting heavy objects may explain why the incidence decreases in later life.

When the prolapse is in a posterolateral direction, the nucleus usually exerts pressure on the corresponding spinal nerve or nerves (see Figs. 2–7, B and 2–7, C). This is because the intervertebral disk forms part of the anterior margin of the intervertebral foramen. Simple decrease in the height of an intervertebral disk also can cause spinal nerve compression. According to Crabbe,[5] in herniation of the disk between L. V. 4 and L. V. 5, the fifth lumbar and/or the first sacral nerve roots are damaged (fourth lumbar root compression is less common). Thus, structures innervated by nerves carrying fibers from these segments will be affected. These nerves (all of which are derived from the sacral plexus) are (1) the *superior gluteal nerve* ($L_{4\text{-}5}$; S_1); (2) the *inferior gluteal nerve* (L_5; $S_{1\text{-}2}$); (3) the *nerve to the quadratus femoris* ($L_{4\text{-}5}$; S_1); (4) the *nerve to*

Fig. 2–7. – **A,** a "typical" vertebra viewed from above. **B,** a thoracic vertebra viewed from the side. The *inferior vertebral notch* combines with the superior vertebral notch of the vertebra below to form an intervertebral foramen for exit of a spinal nerve. **C,** *posterolateral herniation of the nucleus pulposus* of an intervertebral disk, causing pressure on the corresponding *spinal nerve,* in this example on the left side.

the obturator internus (L_5; S_{1-2}); (5) the *posterior femoral cutaneous nerve* (S_{1-3}); (6) the *nerve to the piriformis* (S_{1-2}) and (7) the *sciatic nerve* (L_{4-5}; S_{1-3}). Thus, we might expect to find pain, loss of cutaneous sensation and/or muscle weakness in the areas supplied by any of these nerves. Since only one nerve root (L_5 or S_1) usually is affected and, since none of the nerves named above carries only one of those roots, one would not expect to observe any area of total anesthesia or of individual muscle paralysis. The course and distribution of the sciatic nerve seem to be singled out most frequently by patients in describing lower limb pain (the symptoms of this condition, known as *sciatica,* were described in the case history, i.e., pain over the gluteal area, over the posterior and lateral aspects of the thigh and calf and over the lateral aspect and dorsum of the foot). Less frequently the pain radiates down the medial aspect of the thigh and leg and onto the medial aspect and dorsum of the foot to the big toe; this is characteristic of pressure on the roots of L_4. Backache results from interruption of the *dorsal primary rami* of the lower lumbar or first sacral spinal nerves (the other symptoms discussed above relate to distribution of the *anterior primary rami* of those nerves).

The rationale for conservative treatment (bed rest and analgesics) is to take all possible strain off the affected disk in order to reduce the inflammatory reaction and to allow healing by fibrosis while at the same time controlling pain. Following this initial stage of healing, exercise is recommended in order to facilitate nutrition of the disk and to strengthen the flexors of the spine (i.e., the psoas major muscle). This treatment improves the individual's posture and tends to decrease lumbar lordosis. If conservative treatment is ineffective in relieving the patient's symptoms, surgical excision of the damaged disk usually is indicated.

CASE 2–9: BACK PAIN

History

A 20-year-old male resident of a southwestern communal organization complained of pain in about the middle of the back but, other than that, he felt reasonably well. This pain became chronic, however, and he soon demonstrated fever and lassitude. In addition, a palpable painful lump became evident on the anteromedial aspect of the right thigh, just inferior to the inguinal ligament. The patient's condition worsened daily, prompting him to eventually hitch a ride to the free clinic in the nearest town some 45 miles away. By this time he was experiencing rather severe back pain and considerable limitation of all spinal movements. Physical examination revealed spasms in the anterior abdominal wall musculature and a vague palpable mass deeper in the abdominal cavity just at the right of the vertebral column. Radiographic examination revealed a narrowing of the intervertebral space between T. V. 12 and L. V. 1 and an abnormally bulging but diffuse shadow along the right side of the lower thoracic and upper lumbar vertebrae. There was a similar, but considerably smaller, shadow on the left side. The patient was treated with chemotherapy and surgery in a local hospital and was prescribed a regimen of adequate diet and bed rest. Recovery was rapid and uneventful.

Questions

From the information given, what is your diagnosis of this patient's problem?

Discuss the anatomy and relationships of the muscles that lie along the vertebral column in the posterior abdominal wall.

How do you explain the lump in the patient's thigh?

Discuss the basic pathology of this disease. (You may need to consult the discussion before answering this question.)

What was the rationale for the treatment prescribed?

Discussion

This young man was suffering from *tuberculosis of the spine (Pott's disease),* specifically of T. V. 12 and L. V. 1, with accompanying abscess that entered the psoas sheaths bilaterally. The abscess was more severe on the right side and thus was able to reach the thigh (see below).

The *psoas major* and the *iliacus muscles* together form the *iliopsoas,* which is the chief flexor of the thigh or, when the thigh is fixed, of the trunk. The psoas major, which is basically cylindrical, arises from the intervertebral disks between T. V. 12 and L. V. 1 and from those between each lumbar vertebra (Fig. 2–8). It also takes origin from the margins of the vertebral bodies, from fibrous arches that bridge the points of exit and entry of nerves and blood vessels to and from the vertebral canal and from the transverse processes of the lumbar vertebrae. The fleshy slips combine to form the main mass of the muscle, which descends along the vertebral column, angling laterally as it approaches the pelvis. The iliacus arises from the superior portion of the iliac fossa, from the ala of the sacrum and from adjacent ligaments. It is somewhat fan-shaped. It joins the psoas major muscle and the two pass posterior to the inguinal ligament to insert primarily on the lesser trochanter of the femur, but also into the capsule of the hip joint. As these muscles pass behind the inguinal ligament, they occupy the so-called *muscular compartment*, which lies lateral to the femoral artery and vein *(vascular compartment)*. A *psoas minor muscle* also may be present. It arises from the bodies of T. V. 12 and L. V. 1, and occupies a position on the anterior surface of the psoas major. It usually is represented mainly as a thin, flat tendon, somewhat less than 1 cm in width. It inserts

Fig. 2–8.—Origin, course and insertion of the *psoas major* and *iliacus* muscles.

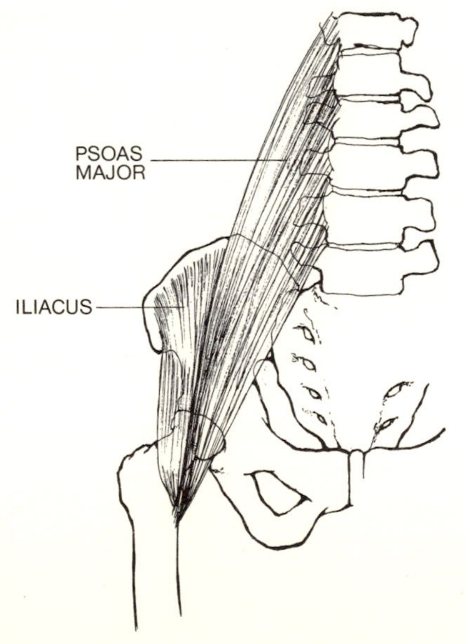

on the arcuate line and iliopectineal eminence of the pelvis and probably assists the psoas major in flexing the trunk. The psoas major and minor are innervated by branches of the lumbar plexus, which is formed in the substance of the former muscle. The iliacus is supplied by the femoral nerve (also a branch of the lumbar plexus), which descends in a groove between this muscle and the psoas major.

The fascia of the posterior abdominal wall is part of the over-all fascial sheath that lines the entire internal surface of the abdominal cavity (sometimes referred to as the *endoabdominal fascia*). This is subdivided into the *transversalis fascia,* which covers the deep surface of the transversus abdominis muscle and the *quadratus lumborum fascia,* which covers the muscle of the same name (this muscle lies just lateral to the psoas major superior to the iliac crest, extending to the twelfth rib), the *iliac fascia,* the *psoas fascia* and so on. The psoas fascia, frequently referred to as the *psoas sheath,* forms a loose investment for that muscle. It is because of this property that an infection, e.g., from a tuberculous abscess of an infected vertebral body, as occurred in this patient, could spread inferiorly along the course of the psoas major muscle and appear near the lesser trochanter. This explains the painful lump in the anteromedial portion of the patient's upper right thigh.

The tubercle bacillus reaches the spine via the bloodstream and attacks the vertebral bodies. According to Crabbe,[5] any portion of the vertebral column may be affected by tuberculosis, "but the commonest site is the thoracic and lumbar spine, particularly at the thoracolumbar junction." Vertebral bodies are affected much more commonly than are elements of the vertebral arch. In our patient, damage to the intervertebral disk produced a narrowing of the disk space that is frequently the first clinically recognizable sign. The disease causes necrosis of the vertebral body, which may eventually collapse, causing deformity and forcing the infected material out into the surrounding tissues. According to Crabbe,[5] "An abscess may burst posteriorly into the vertebral canal but in the majority of instances it tracks forward. In the upper cervical spine a retro-pharyngeal abscess results from anterior rupture* whilst

---

*See also Case 7–10.

in the lower cervical and thoracic spine a para-vertebral abscess occurs. In the lumbar spine the abscess may track postero-laterally to present as a lumbar abscess or it may track down into the pelvis deep to the psoas fascia to produce a psoas abscess." The incidence of tuberculosis of the spine has decreased compared to that of several decades ago, but it is still fairly common, especially in areas where personal hygiene and/or diet are inadequate.

With regard to treatment, appropriate antibiotic drugs are prescribed to eliminate the infectious process, the abscess is drained surgically and an appropriate diet combined with bed rest and local rest is recommended. Psoas abscesses are reached and evacuated via retroperitoneal exposure of the vertebral column. Surgery sometimes also is required to remove necrotic bone segments and to perform bone grafts in attempts to reconstruct deformities of the vertebral column caused by this disease.

CASE 2–10: NECK PAIN AND CONTRACTURE

History

An elderly woman presented to her general practitioner with contracture and pain in the neck. A history revealed that she had been experiencing discomfort in the upper cervical region for several weeks, but it was the muscle spasm and the inability to straighten the neck that prompted her visit to the physician. She was given diathermy treatment and a prescription for a muscle relaxant and analgesics and asked to return in 1 week. The treatment was moderately effective but, as soon as it was relaxed, her symptoms returned and soon became debilitating. She was referred to an orthopedic surgeon who, after a thorough examination, placed her on a 2-week regimen of antibiotics and prescribed bed rest and use of a cervical collar. The symptoms gradually subsided and, after a period of about 8 weeks, she complained of no pain or muscle tightness in the neck.

Questions

What is the name of this condition?

From the information given, what underlying condition do you think was the main contributing factor to the patient's disorder?

Describe the anatomy of the upper cervical vertebrae and how pathology in this area might lead to the observed problem.

Discuss the rationale for the treatment prescribed by each physician.

What are some of the other causes of this condition and how might each be effectively treated?

Discussion

This disorder is known as *torticollis* or wryneck (*tort* = twisted; *colli* = neck; compare with the words tortuous and collar). It can be brought on by numerous conditions (see below). In this case it involved an inflammatory process in the upper two cervical vertebrae, the *atlas* (C. V. 1) and *axis* (C. V. 2), that in turn was the underlying cause of the muscle spasm.

The atlas (the word is derived from the Greek god of the same name whose role was to hold up the world) is unusual in that it possesses neither a body nor a spinous process. It is composed of two lateral masses connected by an *anterior* and a *posterior arch* (Fig. 2–9). The lateral masses articulate with the occipital condyles of the skull above and with the articular facets of the axis inferiorly. *Transverse processes* extend laterally from each lateral mass; each possesses a *foramen trans-*

Fig. 2–9.—The *atlas* (C. V. 1) and its relationship to the *dens* (odontoid process) of the axis (C. V. 2). (Redrawn from Clemente.[6])

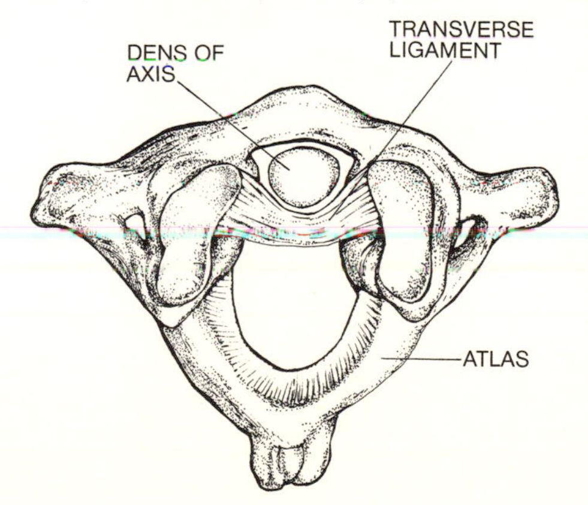

versarium for transmission of the vertebral arteries to the cranial cavity. The anterior arch presents a depression on its posterior surface, the *fovea dentis,* for articulation with the dens or odontoid process of the axis. The *transverse ligament* extends between the internal surfaces of the lateral masses and maintains the position of the dens against the fovea dentis. This joint allows rotary motion around a vertical axis through the center of the dens, hence the name "axis" (or *epistropheus,* which means pivot) for C. V. 2.

The axis, as mentioned above, presents the *dens* or *odontoid process,* which is its most characteristic feature. The inferior parts of the axis begin to resemble those of the more typical cervical vertebrae. Embryologically, the dens actually represents the centrum of the atlas, which separates from it and fuses with the superior aspect of the body of the axis. The tip of the dens is anchored to the anterior margin of the foramen magnum by the *apical ligament;* it also is anchored to the lateral margins of the foramen magnum via the *alar ligaments.* The axis presents articular surfaces for the atlas and C. V. 3 and a thick, bifid spinous process, as well as small transverse processes containing foramina transversaria.

The *rectus capitis lateralis* and the *obliquus capitis superior muscles* attach to the superior aspect of the transverse process of the atlas. The *rectus capitis anterior* has an attachment to the anterior arch of this vertebra, and the *rectus capitis posterior minor* to its posterior arch. These small muscles are concerned with maintenance of posture and movements of the head. Other muscles that attach to the atlas and/or axis are the *levator scapulae, scalenus medius, splenius cervicis, rectus capitis posterior major, trapezius* via the ligamentum nuchae, and some of the small, intrinsic deep muscle bundles of the spine.

Acute torticollis, as seen in this patient, is found mainly in adults and is the result of muscle spasm brought on by an inflammatory or infective process such as tuberculosis of the spine (see Case 2–9) or pyogenic osteitis. This condition is less frequent in children and, when observed, often is related to trauma to the neck or to an infective process in the posterior wall of the nasopharynx (e.g., in the nasopharyngeal tonsil; see Case 7–10). According to Crabbe,[5] "The mechanism of production is inflammatory stretching of the ligaments; particularly the transverse ligament . . . which allows subluxation of one vertebra upon another to occur." As noted above, the inflammation also throws the surrounding muscles into contracture, resulting in wryneck with its associated pain. The rationale for treatment is as follows: (1) Analgesics to reduce pain. (2) Diathermy to warm the tissues and thereby increase circulation to the affected area. Diathermy consists of the use of short wavelength electromagnetic radiation, which, due to resistance at various tissue interfaces (i.e., between skin and subcutaneous tissue or between subcutaneous tissue and muscle), produces localized deep heat. (3) Muscle relaxants to reduce the spasms and contracture. (4) Antibiotics to control and eliminate the infectious process. (5) A cervical collar, which reduces the subluxation of atlas on axis while allowing healing to occur.

Another type of torticollis is referred to as *chronic* (congenital). This is seen in infants as the result of a birth injury, or it may develop intrauterinely. It consists of a contracture of the *sternomastoid muscle* of one side resulting from hematoma and eventual fibrosis of a portion of the belly of the muscle. The sternomastoid muscle originates on the manubrium of the sternum and from up to one third of the medial portion of the clavicle (it is often listed as the sterno*cleido*mastoid, the *-cleid* referring to its clavicular origin). It inserts on the lateral surface of the mastoid process and the lateral portion of the superior nuchal line of the skull. Working with its partner, it flexes the head. Unilateral contraction, however, results in the head inclining to that side, the face rotating to the opposite side and the chin moving superiorly. This is the attitude of the head observed in individuals with chronic torticollis. This condition can be treated in very young infants by gentle stretching of the muscle but, if the problem becomes severe or resistant to manipulation, surgical intervention (resection of the sternomastoid muscle belly) is necessary. Sometimes the condition is severe enough to require total excision of the muscle. This type of torticollis also can occur in adults due to pressure or irritation on the accessory nerve. In this case it is referred to as *neurogenic*.

Other forms of torticollis are spasmodic (intermittent), in

walked into his house, the sense of being surrounded by a crowd of people who loved him. Morey wondered what had happened to those crowds. The dead he'd seen recently enough, but where were the living? Through the years they had scattered and lost touch. Perhaps they had forgotten him. He turned away from the window. Before him were rows of the living, bent over newspapers or simply bent, staring at the floor, worn by the day's work from anxiety to exhaustion.

Morey stood on his front porch. He was only dimly aware of having gotten off the train and driven home, which was often the case. It was as if his brain switched on the parts of him it needed for fine-motor coordination and decision-making and switched off the rest. The ghosts had slipped out of his mind again, but now he remembered. He hesitated.

He turned the key and opened the door. He saw no ghosts, but an unfamiliar shape on his living room coffee table caught his eye: a round, neatly frosted cake. He closed the door behind him.

The cake held more candles than he would have thought it could hold, so many he could barely read the frosting beneath them.

Happy 50th Morey Best!

Morey stood over the cake, letting his keys and briefcase slip from his fingers onto the sofa. The candles burst into flame, and Morey's breath caught in his throat. He looked around the living room, but the dead did not appear. Morey stared into the flames, which dissolved into vague, wavering shapes as his eyes filled with tears. He closed his eyes and leaned forward to blow the candles out.

"I wish—"

which the afflicted individual demonstrates a nervous tic (this type also is referred to as *hysterical torticollis*). It is manifested as a lateral jerking motion of the head, often associated with contraction or contortion of various components of the facial musculature. It usually decreases upon lying down or relaxing but may require psychiatric assistance in its control or elimination. Torticollis also may be brought about by irritation of the inner ear (labyrinthine); it is treated, as in the acute form, by elimination of the irritation. A more common form, and one that is transient, is *myogenic torticollis*, which results from rheumatic or viral (cold) causes or from drafts or, as many patients will suggest when describing their stiff neck, "I must have slept wrong."

CASE 2–11: KNEE INJURY

History

The quarterback of a professional football team suffered a severe knee injury on an unsuccessful pass play. Specifically, he was fading back for the pass when his protection broke down, allowing penetration of two blitzing linebackers. At the time of impact, the right foot was firmly planted on the natural turf, and the right arm was cocked to throw. One of the linebackers hit the quarterback's right knee directly from the lateral side with the full force of his body while the other smashed into the left shoulder, spinning him around in a counterclockwise direction. A stretcher was required to remove the injured player from the field. Upon examination, he complained of great pain in the knee, primarily along the medial aspect of the joint. This pain was accentuated when an abduction or valgus strain (one applied to the lateral side of the injured knee) was exerted by the examining physician. In this manipulation the physician noticed a greater degree of laxity in the patient's right (injured) knee than in the left. There was obvious swelling of the knee, and appropriate tests revealed a large degree of effusion of fluid into the joint cavity. The player also demonstrated a positive McMurray test. In this test the physician placed the index finger and thumb of his right hand along the lateral and medial joint lines, respectively (between femur and tibia), of the patient's flexed right knee, while gently rotating the tibia by twisting the heel one way, and then the other. During this manipulation the physician felt "clicking" sensations deep to his index finger. In addition, a "drawer" test was positive. In this case, the physician noted a greater than normal degree of anteroposterior excursion of the tibia on the femur during appropriate manipulative tests of the knee joint in this plane. The patient was scheduled for surgery, after which the right lower limb was placed in a plaster cast from midthigh to ankle for 8 weeks. This was followed by physiotherapeutic rehabilitation, e.g., exercises designed to strengthen the quadriceps femoris muscle. The player was necessarily out of action for the remainder of the season but was able to return the following year. He demonstrated evidence of right knee joint weakness for the remainder of his career, however.

Questions

From the information given, what structures in the patient's right knee were injured?

Describe the basic anatomy of the knee joint, including the bones, cartilages, ligaments and important muscles associated with this structure as well as the type and action of this joint.

Describe from a biomechanical standpoint how the injury specifically caused damage to the various structures involved.

What was the surgical intervention designed to accomplish?

Would you expect greater than normal laxity of the knee joint postoperatively? Why?

Discussion

This is an example of the classic knee injury that besets football players and is, in fact, responsible for the termination of more football careers than any other single traumatic experience. The fact that the quarterback's cleated shoe was planted firmly in the turf precluded the possibility of the foot sliding at the time of impact to the lateral side of the knee. As a result of this blow, something had to "give" and it was the *medial collateral ligament* of the right knee. Had the player been hit at midthigh level or higher or at midleg level or lower—or had he been playing on artificial turf, which has a greater tendency to allow the foot to slide—the injury might have been avoided. In addition to the blow to the knee, however, he was spun around, i.e., his body rotated counterclockwise by a secondary hit while the right foot was planted. This resulted in the tearing of the *lateral meniscus* and rupture of the *anterior cruciate ligament* of that same knee (see below).

Three bones contribute to the formation of the knee joint, the *femur, tibia* and *patella.* The patella is a sesamoid bone, which occupies a position within the tendon of the quadriceps femoris muscle and articulates respectively with the anterior or inferior surface of the distal end of the femur, depending on whether the knee is extended or flexed. The two tibiae are parallel to one another, but the right and left femora are set at an angle, converging distally (inferiorly). Both the femur and the tibia present *medial* and *lateral condyles,* the respective condyles of the two bones articulating with each other. The fibula does not have any direct association with the femur but, rather, articulates proximally with the tibia only, acting primarily as a strut, with various muscles and ligaments arising or inserting on it. The condyles of the tibia are separated by a nonarticular surface, which is subdivided into an *anterior* and a *posterior intercondylar area.* The central portion of this area presents *medial* and *lateral tubercles* (which actually mark the junction between articular and nonarticular surfaces); together, these constitute the *intercondylar eminence,* which tends to restrict side-to-side displacement of the femur on the tibia. The knee joint contains a fibrous *capsule,* which is attached to the margins of the femur and tibia and which is lined by a synovial membrane.

The concavity of the tibial condyles is deepened by the *medial* and *lateral menisci* (semilunar cartilages). A meniscus, as the student will recall from undergraduate chemistry, is a concave surface (in the physiochemical sense, the interface between liquid and air). The menisci are attached peripherally to the fibrous capsule, the portion between them and the margin of the tibia sometimes being referred to as the coronary ("crown") ligament. The lateral meniscus is shaped somewhat like a doughnut, whereas the medial one resembles more the letter "C" or a rounded horseshoe (Fig. 2–10, A). The central portions are thus open or deficient, allowing the articular cartilage of the femoral and tibial condyles to be separated at these points only by a thin film of synovial fluid. The medial portion of the lateral meniscus and the lateral aspect of the medial meniscus are anchored to the intercondylar area of the tibia by short ligaments. A piece of a meniscus floating free in the joint cavity, which may or may not interfere with normal joint activity, is sometimes referred to as a "joint mouse."

The *medial (tibial) collateral ligament* is a thickening of the fibrous joint capsule, as is the ligamentum patellae. The former extends from the medial aspect of the femur to the medial border of the tibia and is subdivided into superficial and deep bands by the tendon of the semimembranous muscle. The superficial band is crossed on its external (medial) surface by the tendons of the sartorius, gracilis and semitendinosus muscles. The collateral ligaments also resist side-to-side movements of the femur on the tibia. The *lateral (fibular) collateral ligament* is separate from the joint capsule. The *anterior* and *posterior cruciate ligaments*, as their name implies, form an X when viewed from the side. They arise from the anterior and posterior portions, respectively, of the intercondylar area of the tibia, hence their positional classification. The anterior cruciate ligament is attached superiorly to the lateral femoral condyle, whereas the posterior ligament is attached to the medial femoral condyle. The anterior cruciate ligament resists posterior displacement of the femur on the tibia as well as hyperextension of the knee joint. The posterior ligament prevents anterior displacement of the femur or posterior displacement of the tibia. Both resist side-to-side displacement of the femur on the tibia.

The knee joint is strengthened anteriorly and on the sides by muscular or tendinous contributions from the quadriceps femoris, sartorius, semimembranosus and biceps femoris muscles and by the iliotibial tract. Posteriorly, it is reinforced by the popliteus muscle and by the oblique popliteal ligament. It is a *hinge joint*, allowing basically only the actions of flexion and extension, although a slight degree of rotation around a vertical axis also is possible.

The blow to the side of the player's knee caused the following events to occur (Fig. 2–10, B): (1) it widened the normal angle between the femur and the tibia, thus rupturing the medial collateral ligament; and (2) it created an increased vertical force to the lateral aspect of the knee by driving the lateral femoral condyle into the lateral meniscus. The simultaneously induced rotatory force to the left shoulder caused (1) a

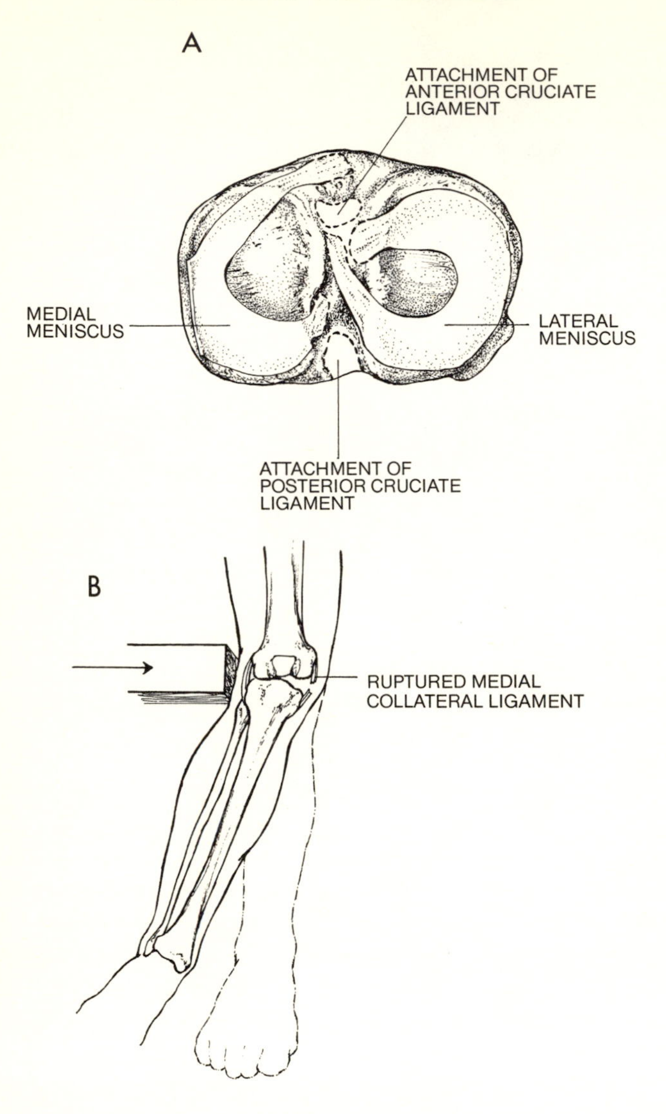
A
ATTACHMENT OF ANTERIOR CRUCIATE LIGAMENT
MEDIAL MENISCUS
LATERAL MENISCUS
ATTACHMENT OF POSTERIOR CRUCIATE LIGAMENT
B
RUPTURED MEDIAL COLLATERAL LIGAMENT

grinding force to be exerted between the lateral condyle of the femur and the lateral meniscus; and (2) a shearing force to be exerted upon the anterior cruciate ligament which, as will be recalled, attaches to the lateral femoral condyle. The result of the combined forces was to tear the lateral meniscus and rupture the anterior cruciate ligament. Surgery thus was required to suture the medial collateral and anterior cruciate ligaments and to remove the damaged lateral meniscus (meniscectomy). Sometimes transplants of other ligaments or tendons, the details of which are not within the scope of this book, are required to reinforce the damaged knee joint. It also should be noted that, if the medial collateral ligament "holds" in injuries of this type, the lateral tibial condyle may be crushed. Meniscectomy increases joint laxity and reduces stability because it decreases the normal degree of concavity between femoral and tibial condyles. If the menisci escape injury, repair of the torn ligaments may restore knee joint stability to near normal.

CASE 2–12: BACK INJURY

History

A 26-year-old male manual worker applied for an unskilled position with a large copper mining corporation. A medical questionnaire, required as part of the application process, revealed that he had injured his back some 5 years earlier in a minor automobile accident. The injury was never diagnosed or treated and it "pretty much went away by itself," although the young man complained of occasional low back pain thereafter. A subsequent medical examination

---

Fig. 2–10.—**A,** the superior articular surface of the tibia. The *menisci* serve to deepen the surface for articulation with the femoral condyles. The *anterior* and *posterior cruciate ligaments* resist anteroposterior dislocation of the knee joint as well as rotary (shearing) forces. **B,** a force to the lateral surface of the knee will result in tearing of the *medial collateral ligament*. The lateral meniscus also may be damaged by compression in this maneuver. Simultaneously applied rotary forces will further damage the meniscus and also may tear the anterior cruciate ligament (redrawn from Rosse and Clawson[4]).

revealed slight defects (nonosseous gaps) in the pedicles of the fifth lumbar vertebra. As a result of this observation, the patient was not offered the position with the mining company but, instead, found a job lifting boxes for a warehouse storage company that did not require a physical examination. Within a few months, however, the back pain became chronic, and a degree of physical deformity was noticeable in the lower spine (increased lumbar lordosis). The range of spinal movements did not seem to be impaired, except perhaps forward flexion; both this action and that of extension were painful. The patient soon became unable to lift anything without experiencing pain and, as a result, consulted a physician. He was placed in a plaster body cast, and a regimen of rest followed by physiotherapy and the use of a firm corset were prescribed. He was able to return to work after a period of 3 months, but within a short time the symptoms returned and persisted. Surgery was therefore performed and subsequently the patient was unable to return to work for nearly a year. However, following his eventual return he experienced no recurrence of symptoms.

Questions

What are the names of the condition in which there is a defect in the vertebral pedicles and of the subsequent and related condition that resulted in physical deformity and pain?

What actually occurs in the latter disorder?

Are these congenital or acquired conditions?

Describe briefly the development (both pre- and postnatal) of a typical vertebra.

Aside from localized pain, would you expect any additional neurologic signs as a result of this disorder? Explain the rationale for your answer.

What do you suppose was the surgical treatment of choice in this case, once it became obvious that conservative management would be ineffective?

Discussion

The defect in the pedicles is known as *spondylolysis*. Much controversy exists regarding the exact cause and nature of this disorder. Until recently it had been considered by a number of investigators to be the result of a congenital lack of or faulty fusion between the vertebral arch and body. Many recent authors,[5,7] however, cite evidence contradictory to that opinion, e.g., the observations that the defect does not occur at the position where normal embryologic fusion does and that defects of this nature are not demonstrable in fetuses or neonates. The incidence of spondylolysis varies considerably with regard to race and age. For example, it is more common in Eskimos than in other groups; some investigators state that it increases in incidence up to age 20, whereas others report it as occurring most frequently in the fourth and fifth decades of life. Most sources agree, however, that it results from a defect in the neural arch in the area between the superior and inferior articular facets. This nonosseous area may contain fibrocartilage.[4] In oblique radiograms of the lower spine, structures of the vertebral arches produce the image of a "Scotty dog," well known to radiologists (Fig. 2–11, B). When spondylolysis is present, the defect in the pedicle presents itself as a "collar" on this dog.

The secondary condition, in which the body of the vertebra (usually L. V. 5) separates completely from the vertebral arch and slides anteriorly on the next inferior vertebra, is known as *spondylolisthesis* (Fig. 2–11, A). Warwick and Williams[8] state that this condition is found in 5% of skeletons. According to most sources, spondylolisthesis is probably preceded in the majority of cases by a defect (i.e., spondylolysis), whether detected or not and whether congenital or acquired (i.e., from an injury, as probably occurred in this case). Once the defect was present in this individual, additional stress (the lifting of heavy boxes) undoubtedly contributed to the displacement of L. V. 5 on the sacrum. This condition, although it seems as though it might be debilitating to the patient, is sometimes asymptomatic and rarely causes neurologic symptoms other than localized back pain. The reason for this is probably that, because the arch stays in place (held by various ligaments and bony articulations), the vertebral canal is simply widened and, therefore, no real pressure is exerted on the nerves of the cauda equina (see Case 2–8) inferior to the defect.

Embryologically two major events occur during the fourth week of development that contribute to the formation of the vertebral column: (1) cells from the sclerotome areas of the somites migrate ventromedially to surround the notochord

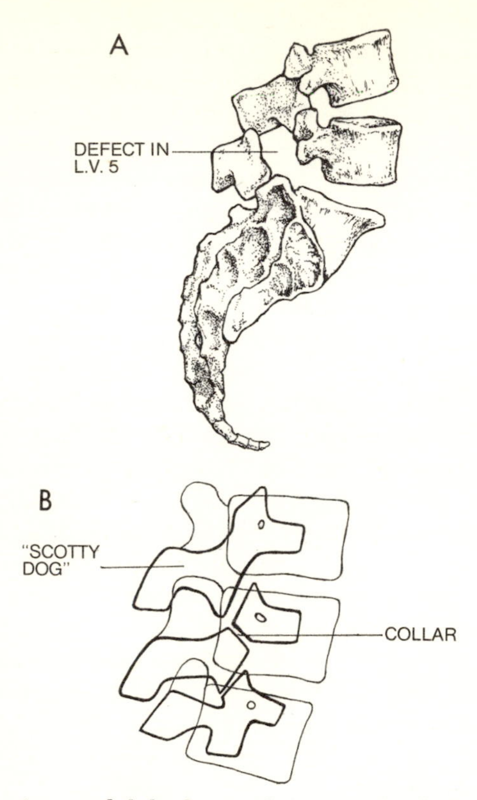

Fig. 2–11. – **A**, spondylolisthesis of *L. V. 5.* The defect in the pedicles has allowed the body of this vertebra to slide forward on the superior articular surface of the sacrum. **B**, oblique view of lumbar vertebrae, which demonstrates the "*Scotty dog*" observed on radiographs. In spondylolysis a "*collar*" on the dog represents the affected vertebral pedicle (redrawn from Crabbe[5]).

and give rise to the *centrum* of the vertebra, and (2) cells from the same area migrate dorsally to surround the neural tube and give rise to the *neural arch* of the vertebra. According to Basmajian,[9] "The body of a vertebra is a composite of the upper and lower epiphyses and the mass of bone between them. It includes the centrum, parts of the neural arch, and the facets for the heads of the ribs. The terms 'body' and 'centrum' are not, therefore, strictly speaking interchangeable; neither are the terms 'vertebral arch' and 'neural arch.' " The central and neural arches chondrify during the sixth week of intrauterine development. Ossification begins during the embryonic period by the appearance of one center in the centrum and one in each half of the vertebral arch. The two halves of the arch fuse during the first neonatal year of life. The arch fuses with the centrum between 3 and 6 years of life. Five secondary ossification centers appear in each vertebra shortly after puberty. There is one for the spinous process, one for each transverse process and one for each of the superior and inferior surfaces of the vertebral body. Ossification of the vertebrae is complete by 25 years of age.

When conservative treatment of spondylolisthesis is ineffective, spinal fusion or ankylosis (of L. V. 5 to the superior aspect of the sacrum in this case) usually is necessary. This may be accomplished by inserting bone grafts between adjacent vertebrae either anteriorly or posteriorly. The latter approach usually is safer and more efficient because it avoids opening the peritoneal cavity, and the exposure is often more complete. In spondylolisthesis of L. V. 5, the graft usually spans the gap from L. V. 4 to the sacrum and thereby "bypasses" L. V. 5.[5]

CASE 2–13: FRACTURE OF THE CLAVICLE

History

The quarterback of a college football team was tackled in such a way that he landed on knees and elbows, with the majority of the weight on the right elbow. While he was in this position, another defensive player landed with his full weight on the quarterback's back. This resulted in a snapped right clavicle for the quarterback. The broken bone was set, and the shoulder was immobilized in a plaster cast for 8 weeks. The player suffered no after-effects of this injury and was able to return to action the following season.

Questions

Briefly describe the bones that constitute the shoulder girdle, specifically those aspects that take part directly in the formation of the shoulder joint.

FIVE O'CLOCK AT THE CARNEGIE LIBRARY

Becca Borawski Jenkins

Every Wednesday, Monica sat on the library steps waiting for her father. Every Wednesday, he failed to arrive. He'd always forgotten. Well, not always, but more times than not. She knew he forgot, so at some point it had ceased to frighten her, no longer made her feel less than or unloved. That's just how her father was—his mind *there* instead of *here*. His body was a misbehaving vessel for a spirit that traveled elsewhere.

Every Wednesday, Monica sat with her stack of books. The agreement was that she'd walk from her school to the library, return her old books and select new ones, and then her father would pick her up on his way home from work. Every week she exchanged at least a dozen books. The librarian doubted that she had actually read them, so Monica dared the woman to ask her a question about each.

"What is the name of Alice in Wonderland's cat?" the librarian asked.

"Dinah," Monica replied.

"What are the names of James and the Giant Peach's aunts?"

where he'd gone wrong. His morning terror was fading fast, leaving nothing but the stink of cold sweat. It was impossible to think of ghosts while trying to predict what his employer would spend next year on products that hadn't been developed yet.

That afternoon, a spatter of applause woke Morey from his computer-induced fog.

"What's that?" he asked no one in particular.

"Some guy from HR's retiring," said a young man—not Nathan, someone five or ten years older.

A moment later, a sheet cake floated past Morey's desk, suspended on the arms of two men in business suits. It disappeared into the break room.

"How long was he here?" Nathan asked the man who apparently knew what was going on.

"Twenty years."

I've got him beat by five, Morey thought.

Nathan smacked his gum. "Jesus. I hope I'm not still here in twenty years."

In the metro station after work, Morey remembered the ghosts. A red line train stopped in front of him, and through its window he glimpsed a girl so pale he thought she was one of his long-lost recent visitors. He stared at her in such obvious terror that she turned her head to stare back as the train pulled away, by which time Morey had realized his mistake.

The train to Vienna opened its doors, and he took a window seat. He leaned against the glass, letting his head be knocked and jostled by the motion of the train. Nathan's comment came back to him suddenly, and a sour taste rose from Morey's throat and collected behind his teeth. For an instant, he had imagined that the applause and cake were for him, although of course no one at work knew it was his birthday. He looked into his face, reflected by the dark window. Maybe you could count wrinkles to figure out someone's age, the way you could count rings on a tree. Memories of his early birthdays drifted through his mind in a pink, nostalgic light. His mother had always planned surprise parties. Some neighborhood kid always gave the surprise away, but it didn't matter. What mattered was the collective shout when he

Discuss in general terms the various joints of the shoulder girdle and the associated ligaments that connect the upper limb to the trunk. As this is done, attempt to explain why the clavicle was broken rather than the shoulder (or some portion thereof) dislocated.

Discussion

The *shoulder (shoulder girdle)* is composed of the *clavicle, scapula* and *proximal humerus* and the various capsules, ligaments and muscles that connect them to each other. The overall structure is, in effect, the means of connecting the upper limb to the trunk and, as such, it combines a maximum of mobility with a maximum of strength. The shoulder girdle is commonly subdivided into four separate joints *(glenohumeral, acromioclavicular, sternoclavicular* and *scapulothoracic).* The so-called "shoulder joint" is represented to most people, however, by the glenohumeral portion of the overall shoulder girdle.

The *clavicle (collar bone)* is S shaped in the horizontal plane. It extends from the superolateral angle of the manubrium to the acromion of the scapula. Its medial extent is rounded, whereas it becomes flattened laterally in the horizontal plane. The clavicle serves as a strut between the tip of the shoulder and the sternum. The inferior surface of the lateral third of the bone presents the *conoid tubercle* and the *trapezoid line* for the attachments of the ligaments of the same names, which together constitute the *coracoclavicular ligament.*

The *scapula (shoulder blade)* is a complex bone, all aspects of which do not concern us here. The portions that take part in the formation of the shoulder girdle are: (1) the *acromion,* which is the lateral extension of the spine of the scapula and which forms the prominent bony "point" of the shoulder; (2) the *glenoid cavity* or *fossa,* which is an anterolateral extension of the body of the bone and which receives the head of the humerus; and (3) the *coracoid process,* a fingerlike structure that projects anterolaterally from the superior border of the scapula (Fig. 2–12).

The proximal end of the *humerus* presents a hemispherical

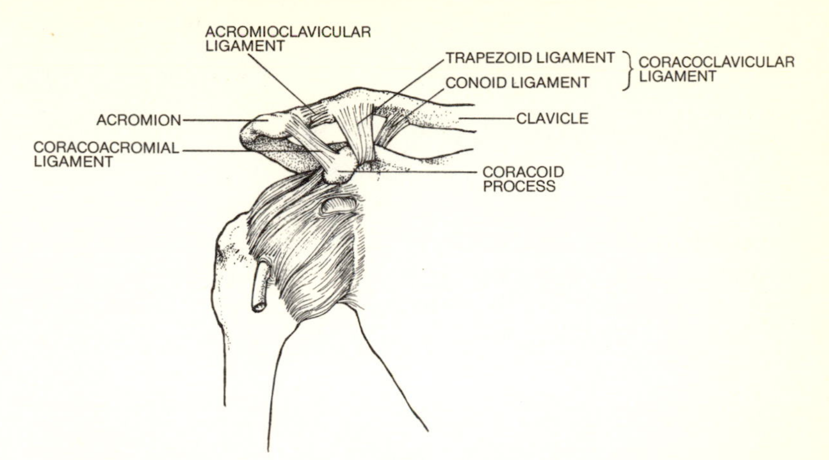

Fig. 2–12. – The *coracoclavicular (conoid* and *trapezoid) ligament.* A powerful force directed vertically upward through the humerus may result in a broken clavicle rather than in rupture of this strong ligament.

head, which articulates with the glenoid cavity and which is separated from the remainder of the bone by the *anatomical neck.* Opposite the head the bone presents a lateral *greater tuberosity* and a medial *lesser tuberosity,* which are separated by the *intertubercular* or *bicipital groove,* which in life houses the tendon of the long head of the biceps brachii muscle. The *surgical neck* of the humerus is the portion between the proximal end and the shaft, a site of frequent fractures of this bone.

The *glenohumeral joint,* "universal" in function, permits the widest range of motion of any joint in the body. It is not a true ball and socket joint, as is the hip joint, because of the shallowness of the glenoid cavity (the latter is deepened somewhat, however, by a cartilagenous lip, the *glenoid labrum*). The capsule of this joint extends from the margin of the glenoid cavity to the anatomical neck of the humerus; medially it extends farther down the shaft of the humerus than elsewhere, which allows full abduction of the upper limb. The joint capsule is strengthened somewhat by the *coracohumeral ligament* and by the often difficult-to-demonstrate superior, middle and inferior *glenohumeral ligaments.* The main strength of the joint, however, is not a result of the bony arrangement (which, indeed, offers almost no stability at all) or of these ligaments but, rather, a result of the *rotator (musculotendinous) cuff.* The latter is derived from the tendons of the *supraspinatus, infraspinatus* and *teres minor muscles,* which cross the joint and insert on the greater tuberosity of the humerus, and from the tendon of the *subscapularis muscle,* which inserts on the lesser tuberosity. This cuff greatly strengthens the joint, primarily anteriorly, but it is incomplete inferiorly; and dislocations (subluxations) occur more frequently at this joint than at any other joint of the body.

The *acromioclavicular joint* is a plane joint between the lateral end of the clavicle and the medial border of the acromion. It possesses a short, tight capsule (often described as having specific thickenings, the *acromioclavicular ligaments*), and the range of movement at this joint is limited. The main ligamentous strength of this joint is derived from the *coracoclavicular ligament,* which as its name implies extends vertically upward from the coracoid process to the inferior surface of the clavicle. This ligament is subdivided into the medially placed *conoid ligament* and the more laterally placed *trapezoid ligament.* The two are shaped as their names imply, the conoid ligament resisting vertically directed forces, and the trapezoid ligament the more horizontally directed shearing type of forces. The *coracoacromial ligament* extends from the lateral margin of the coracoid process to the acromion and therefore does not cross any joints per se (i.e., it extends between two processes of the same bone). It does, however, form a protective dome over the head of the humerus.

The *sternoclavicular joint* is formed by the association between the rounded medial end of the clavicle and the manubrium and first costal cartilage. A joint capsule is strengthened by *anterior* and *posterior sternoclavicular ligaments.* An *interclavicular ligament,* which spans the suprasternal (jugular) notch, and *costoclavicular ligaments* also may be defined. The joint cavity is divided into medial and lateral halves by a fibrous or fibrocartilaginous *articular disk.* The *scapulothoracic joint* is simply that association between the costal surface of the scapula and the posterolateral aspect of the thoracic wall. It allows gliding movements and rotary movements (often likened to the action of a wing nut). The combined motion of the scapula and humerus, which for one thing allows abduction of the upper limb above the horizontal plane, is known as *scapulohumeral rhythm.*

A strong blow to the lateral aspect of the shoulder may drive the acromion under the lateral end of the clavicle, resulting in a *shoulder separation* (dislocation). Medial movement of the clavicle is resisted in such incidents due to the stability of the sternoclavicular joint. The acromioclavicular joint is more likely to give way than is the clavicle likely to break in such blows, since the force is directed parallel to the shaft of the clavicle. This type of dislocation also may occur from a fall on the outstretched hands. A fall on the elbow, as occurred in this case, especially when weight was applied vertically, resulted in a force being directed parallel to the fibers of the coracoclavicular ligament, especially the conoid portion thereof. The ligament proved to be stronger than the bone, resulting in a broken clavicle. Rupture of the coracoclavicular ligament would have resulted in separation of the acromioclavicular joint.

It is interesting to note that, although the cat possesses clavicles, they do not articulate with the scapulae but rather are relatively rudimentary, being simply embedded in muscle. Likewise, the scapulae are embedded in muscle and have no direct bony connection to the thoracic cage of the animal. As a result, the cat can land with great force on its front feet without any danger of a broken clavicle or shoulder separation. The next time you happen to see a cat jump from a high place, observe how the scapulae are forced dorsally upon impact and how the various muscles and ligaments operate as shock absorbers in this maneuver.

CASE 2–14: UPPER LEG INJURY

History

A foot soldier was struck by a piece of shrapnel from a hand grenade at the upper lateral surface of the left leg, just below the knee.

Specifically, the wound, which did not appear severe, occurred just below the lateral surface of the head of the fibula. Upon inspection and after cleaning, the neck of the bone was visible through the broken skin. The wound was sutured by a medic, and the soldier was sent back to the front lines, regardless of his protests that the action of the left foot had become clumsy: he found that he tripped rather consistently over ordinarily avoidable objects. During the ordinary swing phase of locomotion, he had difficulty keeping the toes from scraping the ground unless flexion at the thigh was greater than normal to compensate for this (the so-called "steppage" gait). When the foot approached the ground at the beginning of the stance phase, it slapped the ground, acting more like a flipper than a foot. The soldier also could no longer evert the foot and had a tendency when stepping on small rocks or similar items to lose stability of the ankle by hitting on the lateral side of the foot. The resultant mechanically produced inversion caused him to repeatedly sprain the left ankle. His sergeant, recognizing the problem, sent him to the Red Cross hospital, where the injury was considered more serious than originally thought. In addition to the postural and gait difficulties, neurologic examination revealed loss of cutaneous sensation over the lateral aspect of the injured leg and over the dorsum of the foot. The injury resulted in a medical discharge from the Army and in a permanent disability for the young man.

Questions

What specific nerve was injured, and where? Trace the course of this nerve from its origin to its terminal branches.

Discuss the soldier's symptoms on the basis of the muscles and cutaneous areas supplied by the branches of this nerve.

How susceptible do you think this nerve is to injury, and what is some other way that it might be damaged?

Are prospects for regeneration good, average or poor?

Discussion

The *common peroneal nerve* was severed at the point where it winds laterally around the neck of the fibula. This nerve is very superficial and therefore subject to injury at this point. The student can substantiate this statement by palpating the head of his or her own fibula and then extending the fingers inferiorly to the neck of the bone. With deep pressure at this point, the nerve can often be rolled against the bone at the point where the former structure crosses the latter. Unfortunately for the soldier, even though his wound was basically not severe, this was the exact point at which the piece of shrapnel hit.

The common peroneal nerve (L_{4-5}; S_{1-2}) is one of the two terminal branches of the sciatic nerve, the other being the tibial nerve. It arises at the apex of the popliteal fossa and courses laterally, usually under cover of the medial border of the biceps femoris muscle. It crosses the lateral head of the gastrocnemius, reaches the head of the fibula and then winds superficially around the neck of that bone, as noted above. At this point, it usually divides into its two terminal branches, the *superficial* and *deep peroneal nerves* (Fig. 2–13). Before doing this, however, the common peroneal usually gives rise to the *lateral sural cutaneous nerve,* which provides cutaneous sensation to the lateral portion of the upper leg, extending from about the midanterior surface around approximately to the midposterior surface of the leg. A *peroneal communicating branch* also may arise from the common peroneal. This nerve, when present, joins the medial sural cutaneous branch of the tibial nerve to form the *sural nerve,* which supplies the skin of

Fig. 2–13.—Relationship of the *common peroneal nerve* and its subdivisions, the *superficial* and *deep peroneal nerves,* to the neck of the fibula.

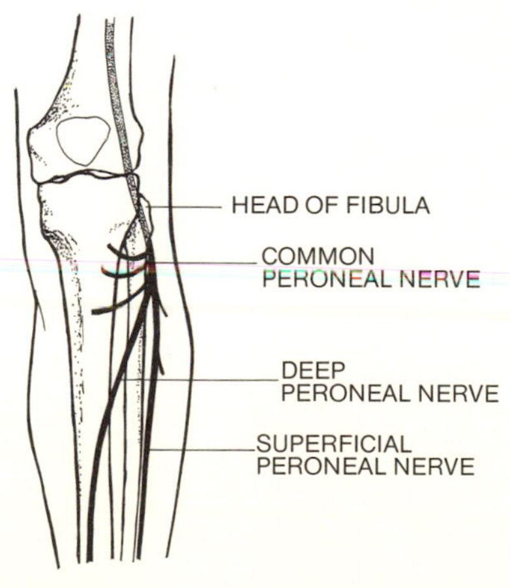

the back of the leg and the lateral surface of the foot. Considerable variation is evident in the origin and communication of these nerves.

The *superficial peroneal (musculocutaneous) nerve* descends along the anterior surface of the fibula between the peroneus longus and brevis muscles, which it supplies, and the extensor digitorum longus muscle. The *peroneus longus* everts the foot and is a plantar flexor. The *peroneus brevis* also everts the foot. The cutaneous branches of the superficial peroneal nerve are the *medial* and *intermediate dorsal cutaneous nerves*. The former supplies the big, second and third toes with cutaneous sensation; the latter supplies this form of innervation to the fourth and fifth toes. There are considerable overlap and variation in this pattern. The dorsal aspects of the tips of the toes are supplied by the medial and lateral plantar nerves, branches of the tibial nerve (compare this to the innervation in the hand, in which the median nerve supplies the dorsal aspects of the tips of certain fingers; see Case 2–5).

The *deep peroneal nerve* pierces the anterior intermuscular septum of the leg after winding anteriorly around the neck of the fibula. It also pierces the *extensor digitorum longus muscle,* which it supplies, and then descends on the interosseous membrane under cover of this muscle and the *extensor hallucis longus*, which it also supplies. It meets the anterior tibial artery on the interosseous membrane and descends with that vessel. It gives muscular branches in the leg also to the *tibialis anterior* muscle and to the *peroneus tertius.* It should be noted that in certain instances the innervation of the peroneus longus, the tibialis anterior and/or the extensor hallucis longus may be derived from the common peroneal nerve in the popliteal fossa. The extensor digitorum longus and extensor hallucis longus muscles have actions that are indicated by their names (e.g., hallux = big toe). The tibialis anterior dorsiflexes and inverts the foot. The peroneus tertius really is the lateral portion of the extensor digitorum longus, although its insertion is into the base of the fifth metacarpal or adjacent fascia. It is a weak evertor of the foot. The deep peroneal nerve also supplies the *extensor digitorum brevis muscle* in the foot (lateral terminal branch). The medial terminal branch supplies *dorsal digital branches* (cutaneous innervation) to a small triangular patch of skin between the big and second toes. It is difficult to determine this loss of cutaneous innervation alone (if, for instance, only the deep peroneal nerve were cut) due to communication and overlap from the cutaneous branches of the superficial peroneal nerve.

The result of section of the common peroneal nerve is primarily manifested by a loss of dorsiflexion and eversion, resulting in *footdrop*, or a foot that dangles in an *equinovarus position*. This is primarily due to denervation of the tibialis anterior (although the long muscles that extend the toes also aid in dorsiflexion of the foot) and of the peroneus longus and brevis. It is obvious that the patient also cannot extend the toes. The slapping of the ground is the result of unopposed action of the plantar flexors, i.e., an inability to synergistically control and thereby decelerate plantar flexion. The sensory losses were noted in the history.

The common peroneal nerve is the most susceptible branch of the sciatic nerve to injury[10] and one of the most susceptible in the body. According to Gardner *et al.*,[3] the nerve has been damaged, apparently by stretching, due to the weight of blankets on the upward-pointing toes of patients confined to bed for long periods of time. It also has been known to be damaged by pressure from plaster casts and in injuries that cause edema or other pressures to be exerted upon the nerve, resulting in necrosis. Regeneration is said to be poorer than that of any other nerve in the body, perhaps due to separation of the nerve ends upon section; that which does occur does so at a very slow rate. A toe-raising device may be used to control footdrop in patients with permanent disability. This consists simply of a spring attached by straps between the toes and calf.

CASE 2–15: FLAT FEET

History

The parents of a two-and-one-half year old boy noticed that he was walking on the insides of his feet. This was made especially evident by the uneven wear pattern of the shoes. The parents had thought earlier that their son had abnormally flat feet prior to walking, but their pediatrician had suggested that they delay further investigation until the boy's walking habits had become fairly well established. An examination at this time revealed that the boy did, in fact, have *flat feet (valgus foot* or *pes planus),* which was having a deleterious effect on his walking. In addition, the child seemed to experience discomfort while walking or running, and he was clumsier than other children of the same age or relative stage of development. He was referred to an orthopedic surgeon who, after a thorough examination that included a radiologic study of the boy's feet, recommended conservative treatment in the form of corrective shoes. He was fitted with shoes that contained a built-up arch system and was instructed to go barefoot as little as possible. This treatment continued for 12 years, after which time the boy demonstrated no evidence of flatfoot. He passed his Army physical examination some years later and was subsequently drafted into that branch of the service.

Questions

What exactly are flat feet?

Very basically describe the bony, ligamentous and muscular anatomy of the feet as it applies to the maintenance of their arched configuration.

Describe the "arches" of the feet.

Explain the rationale for the conservative form of treatment in reducing and eventually eliminating the patient's problem.

Discussion

A great deal of controversy exists as to what conditions of the feet are accurately included in the general term flatfoot. Many investigators feel that the term is unsatisfactory to describe the variety of disorders currently ascribed to it. Technically, a flat foot is simply one in which there is a depression in the normal height of the longitudinal arch. Infants have subcutaneous pads of fat on the plantar surfaces of their feet and, as a result, the newborn's footprint is characteristically flat. This condition changes with age as the fat is redistributed and greatly reduced. In addition, the longitudinal arch normally increases in height throughout childhood, as muscular support of the foot (see below) becomes more prominent. There is considerable variation in what is considered a normal arch, but most investigators and physicians involved in this field believe some concavity of the medial aspect of the foot should be evident (e.g., in footprints). Some individuals possess abnormally high arches *(pes cavus);* their footprints may demonstrate a lack of continuity between the heel print and the remainder of the foot. Conditions that probably should not be included in the ordinary category of flatfoot are congenital vertical talus (mallet foot or rocker bottom foot) and rigid flatfoot (due to an abnormal synostosis or synchondrosis). These conditions almost inevitably require surgical intervention for correction.

The skeleton of the foot consists of seven *tarsal bones,* five *metatarsals* and 14 *phalanges.* The tarsals are the *talus, calcaneus, navicular, cuboid* and three *cuneiform* bones. The body of the talus articulates with the tibia and fibula as the *ankle joint.* The bone also has a neck and a head. The head articulates anteriorly with the navicular and inferiorly with a medial shelf of the calcaneus, the *sustentaculum tali,* as the *subtalar (talocalcaneal) joint.* The calcaneus is the largest of the tarsal bones and forms the prominence of the heel. The anterior end of this bone articulates with the cuboid. The navicular occupies a medial position in the foot, articulating with the talus behind and the three cuneiforms anteriorly. The cuboid is lateral to the navicular and, being longer than that bone, extends between the calcaneus and the fourth and fifth metatarsals. The three cuneiforms are, as their name implies, wedge shaped, with the wide surface directed superiorly, thereby forming a round type of arch in coronal section, with the middle (intermediate) cuneiform acting as the so-called keystone or highest segment. They articulate anteriorly with the first, second and third metatarsals. Each metatarsal presents a base (for articulation with the cuneiforms and cuboid), a shaft and a head. The heads articulate with the bases of the proximal phalanges. Each phalanx also has a base, shaft and head. There are two phalanges in the big toe and three in each of the remaining toes.

The major interosseous joints of the foot are the *subtalar* (see above), the *talocalcaneonavicular* and the *transverse tar-*

the ghost, grabbed yesterday's rumpled work clothes from the floor, locked himself in the bathroom, and dressed. He had to escape, but he wasn't about to run outside half-naked. Just because he was hallucinating dead people didn't mean he was going to give the whole neighborhood something to gossip about. By the time Morey's hands stopped shaking enough to let him button his shirt, his pulse had slowed. Of course there was no ghost in his room! He had woken from a very realistic nightmare, that's all. He took a deep breath and opened the bathroom door.

His mother looked concerned.

Before she could speak, Morey bolted for the front door. Shoes, car keys, briefcase, door, he thought, focusing on the concrete details of his escape. As he ran through the living room, his house erupted with voices.

"HAPPY BIRTHDAY!"

Dead uncles, aunts, grandparents, and childhood friends burst from their hiding spots behind chairs and sofas. Some of them blew noisemakers and threw confetti. Morey stumbled and grabbed his shoes, car keys, and briefcase as he rushed outside, slamming the door behind him. Sweat soaked his white shirt. He tottered a few steps from the house and sat on the sidewalk to put on his shoes. It was too early for Morey to leave for work, but he wanted to be around people, somewhere safe and mundane. On the metro, he looked around to reassure himself that the dead were not among the usual crowd of sweaty, overly cologned suburbanites on their way to jobs in the city.

Morey was the first person in the office. He turned on his computer and opened the spreadsheet where he kept track of who was supplying armaments to whom. Nathan, the cubicle neighbor who chewed gum too loudly, was the next person to arrive. He was decades younger than Morey, which seemed true of more of Morey's coworkers every year.

"Tim told me to tell you there's something wrong with the calculations you sent last night," Nathan said as he passed Morey's desk.

Morey turned to ask for details, but Nathan had already vanished behind the gray carpet wall that separated them. Morey checked his email for sent messages, trying to figure out what he'd told Tim and *sal*. The latter consists of the *talonavicular* portion of the talocalcaneonavicular and *calcaneocuboid joints*. There are also *cuneonavicular, intercuneiform, cuneocuboid, tarsometatarsal, intermetatarsal, metatarsophalangeal* and *interphalangeal joints*. Although these joints possess capsules with numerous named ligamentous thickenings located on both the dorsal and plantar surfaces of the foot, there are three ligaments primarily concerned with maintenance of the arched configuration of the foot (Fig. 2–14): (1) The *plantar calcaneonavicular ("spring") ligament*, which extends from the sustentaculum tali of the calcaneus to the inferior surface of the navicular. The head of the talus fits into the gap between these two bones, resting inferiorly on this ligament. As a result, the weight of the body is transmitted through the talus to the spring ligament, which acts something like a shock absorber in transmitting the forces of the body to the foot. (2) The *long plantar ligament*, which extends between the plantar surface of the calcaneus and the tuberosity of the cuboid and extends to the bases of the third, fourth and fifth metatarsals. (3) The

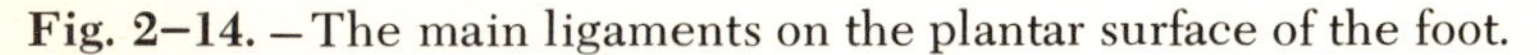
Fig. 2–14. – The main ligaments on the plantar surface of the foot.

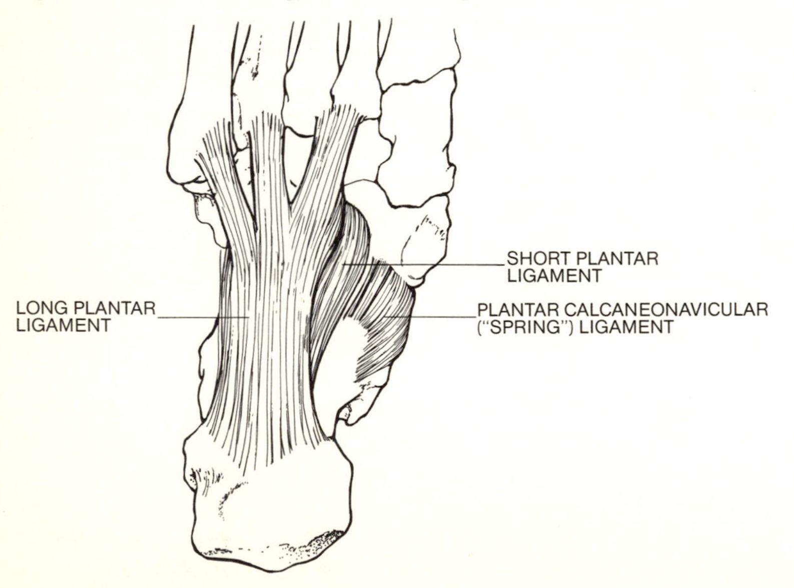

short plantar ligament, which for the most part lies deep (dorsal) to the long plantar ligament. It also extends between the calcaneus and the cuboid.

Most sources agree that the bony configuration and ligamentous ties of the foot are the primary factors in maintaining the arches of this structure. Most agree similarly that the intrinsic muscles of the foot are largely inactive in standing, but come into play during walking or other movements. On the other hand, certain clinicians believe that flatfoot in infants is due primarily to muscular immaturity and that, as these muscles become developed, the arched nature of the feet is accentuated. For example, Crabbe[5] states, "If the adductor hallucis fails the transverse arch is lost and the metatarsal heads 'prolapse.' . . . Any failure of the intrinsic muscles allows the toes to buckle with loss of the transverse arch and 'prolapse' of the metatarsal heads." Certain long muscles also assist in maintaining arch structure. For example, the tibialis anterior exerts a superiorly directed force on the medial cuneiform, and the tibialis posterior and peroneus longus muscles are said to form a sling under the arches of the foot.

The arches of the foot usually are described as *longitudinal* and *transverse.* The longitudinal arch is commonly subdivided into a *medial arch* (composed of the calcaneus, talus, navicular, cuneiforms and medial three metatarsals) and a *lateral arch* (composed of the calcaneus, cuboid and the lateral two metatarsals). The transverse arch is more complicated, consisting actually of a series of arches running transversely across the foot. It is much higher medially than laterally. For this reason the arches of the feet have been likened to half domes, which become a full dome when the medial surfaces of the two feet are placed together.[11] Dorsiflexion and plantar flexion of the foot occur primarily at the ankle joint, although a limited degree of these movements is possible at the subtalar and transverse tarsal joints. The greatest degree of inversion and eversion occurs at the latter two joints. Motion is very limited at the cuneonavicular and tarsometatarsal joints.

Conservative treatment for "ordinary" flatfoot is based on providing artificial support for the arches of the foot by a mechanical device (i.e., orthopedic shoes), while waiting for the intrinsic muscles of the feet to develop enough strength to provide the necessary natural support. This procedure also takes undue stress off the ligaments of the plantar aspect of the foot and allows the bones to develop normally.

CASE 2–16: INJURY TO THE AXILLA

History

A man was propping himself up with an outstretched arm against a tree in his backyard watching his son mow the lawn when a rock thrown by the power mower struck him sharply in the exposed armpit. Once the initial pain subsided, he observed that he could no longer abduct his arm on the injured side above the horizontal plane. Examination revealed that the medial border of the scapula on that side stood away from the posterior aspect of the rib cage. The injury in the armpit did not appear severe (although the skin was broken, the wound was not deep), and within several days he regained the ability to fully abduct the arm. The protrusion on the back subsided and disappeared concomitantly.

Questions

What is the name of this condition and what is its cause?
Describe the anatomy of the structures involved.
Why was this condition transient?

Discussion

This condition is known as *"winged" scapula* and is caused by paralysis of the *serratus anterior muscle,* which is supplied by the *long thoracic nerve* (C_{5-7}). The rock evidently was propelled at sufficient velocity to cause trauma to the nerve (probably by compressing it against a rib). The recovery of function of the man's upper limb and the gradually diminishing protrusion on his back was evidence that the nerve was not cut but only temporarily injured by the rock. Thus, recovery eventually ensued without treatment.

The serratus anterior muscle arises from the external surfaces of the upper 8 ribs and inserts along the entire medial (vertebral) border of the costal surface of the *scapula* from its

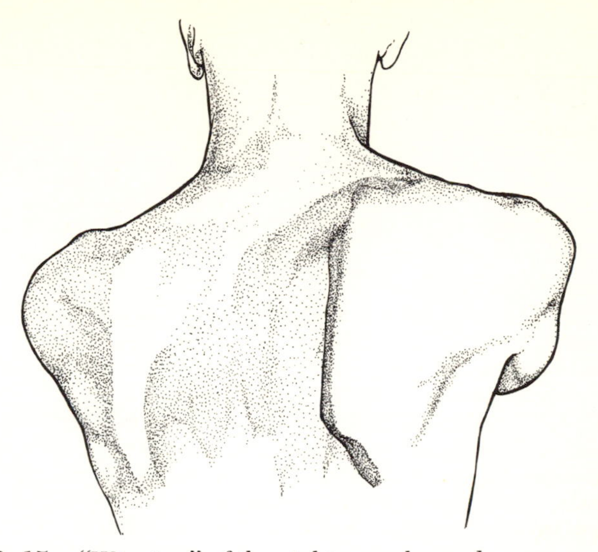

Fig. 2–15. – "Winging" of the right scapula, made more prominent by having the patient press his outstretched hands against a wall. (Based on Spillane, J. D.: *An Atlas of Clinical Neurology* [New York: Oxford University Press 1968].)

superior to its inferior angle. The muscle forms the medial wall of the *axilla.* The serratus anterior maintains the proximity of the medial border of the scapula to the rib cage, hence the protrusion or "winging" caused by its paralysis. The winging becomes especially evident when the patient is asked to push against a wall with his hands (Fig. 2–15). The inferior fibers twist the scapula like a wing nut around an imaginary axis through the center of the bone, thus moving the *glenoid fossa* upward and allowing abduction of the arm above the horizontal plane. When there is paralysis of the serratus anterior, the deltoid muscle is capable of abducting the arm only to approximately the horizontal plane.

The long thoracic nerve is derived from branches of the ventral rami of nerves, which contribute to the brachial plexus. The roots of the nerve combine and descend posterior to the plexus and the axillary artery. The nerve then comes to lie on the external surface of the serratus anterior muscle approximately in the midaxillary plane. It is superficial in this position and is therefore subject to injury, provided the arm is abducted, as was the case in this freak accident.

REFERENCES

1. Tobin, C. E.: *Shearer's Manual of Human Dissection* (5th ed.; New York: McGraw-Hill Book Co., 1967).
2. Marks, C.: *Applied Surgical Anatomy* (Springfield, Ill.: Charles C Thomas, 1972).
3. Gardner, E., *et al.: Anatomy* (3d ed.; Philadelphia: W. B. Saunders Co., 1969).
4. Rosse, C., and Clawson, D. K.: *Introduction to the Musculoskeletal System* (New York: Harper & Row, 1970).
5. Crabbe, W. A.: *Orthopaedics for the Undergraduate* (London: William Heinemann Medical Books, Ltd., 1968).
6. Clemente, C. D.: *Anatomy—A Regional Atlas of the Human Body* (Philadelphia: Lea & Febiger, 1975).
7. Hollinshead, W. H.: *Textbook of Anatomy* (3d ed.; Hagerstown, Md.: Harper & Row, 1974).
8. Warwick, W., and Williams, P. L.: *Gray's Anatomy* (35th British ed.; Philadelphia: W. B. Saunders Co., 1973).
9. Basmajian, J. V.: *Grant's Method of Anatomy* (8th ed.; Baltimore: Williams & Wilkins Co., 1971).
10. Sunderland, S.: The relative susceptibility to injury of the medial and lateral popliteal divisions of the sciatic nerve, Br. J. Surg. 41: 300, 1953.
11. McKenzie, J.: The foot as a half-dome, Br. Med. J. 1:1068, 1955.

Chapter 3

RESPIRATORY SYSTEM

CASE 3–1. LUNG CANCER

History

A 67-year-old male who had been a heavy cigarette smoker for nearly 50 years was diagnosed as having lung cancer. Radiologic and bronchoscopic examinations revealed a relatively small tumor in one of the tertiary bronchi of the middle lobe of the right lung. This segment of the patient's lung was removed and, since there was no evidence of metastasis of the tumor, no further treatment was prescribed. He quit smoking and 5 years after the surgery was free of any detectable cancer.

Questions

Describe the basic anatomy of the lungs and the bronchial tree.

Why was it possible to remove only a portion of the diseased lung rather than a lobe or the entire structure?

Discussion

Each lung presents an *apex*, which ordinarily extends slightly above the level of the first rib and clavicle, and a *base* or *diaphragmatic surface*. The *medial* or *mediastinal surfaces* of the two lungs face each other; the remainder of the lung surface, by far the most extensive, is the *costal portion*. Each lung is enclosed within a double-layered, closed *pleural sac;* the potential space between the *visceral* and *parietal* layers of the pleura is the *pleural cavity*. Each lung presents an *oblique fissure*, which when viewed from the side follows approximately the line of the sixth rib and divides the lung basically into two halves. Thus, the left lung contains two *lobes*, a *supe-*

rior and an *inferior*. An additional fissure, the *horizontal fissure,* is present in the right lung. It follows approximately the line of the fourth rib from the oblique fissure to the sternum. As a result, the right lung is subdivided into *superior, middle* and *inferior lobes*. The portion of the medial surface of the lung where the various structures enter and leave is known as the *hilus*. The structures that enter and leave collectively constitute the *root of the lung*.

The *trachea* extends inferiorly from the cricoid cartilage of the larynx to the level of the sixth thoracic vertebra where it divides into *right* and *left main (primary) bronchi*. Because the trachea inclines to the right before it divides, the angle between the trachea and the right bronchus is greater than that between the trachea and the left bronchus. As a result, inhaled objects are more frequently found in the right lung than in the left. Each main bronchus then divides into the *lobar (secondary) bronchi;* these are termed the same as the lobes of each lung. The lobar bronchi subsequently give rise to the *segmental (tertiary)* bronchi.

About a quarter of a century ago it was recognized that the human lung was divisible for both descriptive and surgical purposes into a number of *bronchopulmonary segments*. Each of these is separated from adjacent segments by connective tissue septa, and each possesses its own segmental bronchus and branch of a pulmonary artery. The veins, on the other hand, are intersegmental in nature. The right lung possesses 10 such segments, and the left lung nine. As a result of the bronchial and arterial segmentation, a disease process may be identified within a single bronchopulmonary segment, and frequently that single segment may be removed, as occurred in this case. For the nomenclature of these segments, the student is encouraged to consult a standard anatomy text.

CASE 3–2: LYMPHATIC DRAINAGE OF THE LUNG

History

An elderly female also (see Case 3–1) was diagnosed as having lung cancer. Unfortunately, the disease had metastasized to the lymphatic system, and her condition was considered inoperable and therefore terminal. She died 5 months after diagnosis.

Questions

Trace the lymphatic drainage (and hence the metastatic spread of tumor cells) from the lung to the systemic circulation.

Discussion

The small lymph nodes associated with the lung tissue itself are known as *pulmonary nodes*. They are commonly the first nodes to become involved in infectious diseases or tumors. They lead to the *bronchopulmonary nodes*, which are located in the vicinity of the hilus of the lung and may become considerably enlarged in disease processes. In the dissecting room these lymph nodes often are seen to contain large quantities of carbon filtered from the air or from tobacco smoke. These two sets of lymph nodes lead to the *tracheobronchial nodes* (Fig. 3–1), which, as their name implies, lie at the bifurcation of the trachea into the two main bronchi. These nodes are divided into an *inferior set*, which lies below the bifurcation of the tra-

Fig. 3–1. – Lymphatic drainage of the lungs and bronchi.

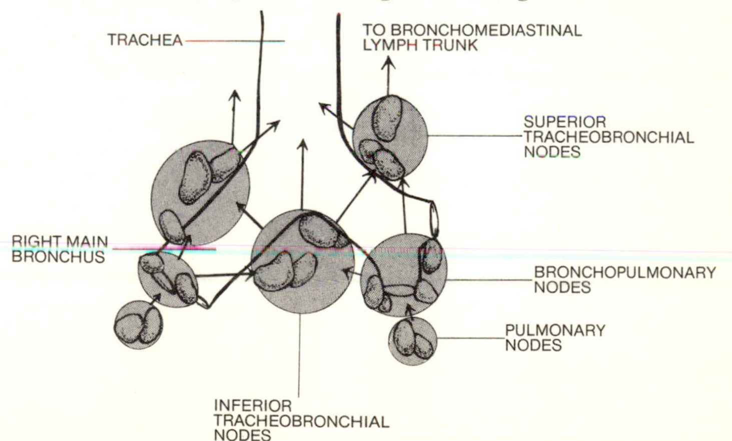

chea, and into right and left *superior sets*, which lie in the angle between the trachea and main bronchus on each side.

The tracheobronchial lymph nodes lead on each side to the *right* and *left bronchomediastinal lymph trunks*, which also receive afferent lymphatic vessels from mediastinal nodes. The left bronchomediastinal trunk drains to the thoracic duct or to one of the major veins on the left side of the neck (internal jugular, subclavian or brachiocephalic). On the right, this trunk either enters the right main lymphatic duct or a corresponding vein on the right side of the neck. Lung cancer is very difficult to eradicate once it has spread to the lymphatics of the thorax due to the difficulty of their total removal and to their close physical association with the general systemic circulation.

CASE 3–3: HISTOLOGY OF THE LUNG

History

A 53-year-old male who had smoked between two and three packs of cigarettes since the age of 14 years had noticed over the previous 2 years a progressive difficulty in ordinary respiration. Specifically, this was manifested as fairly normal inspiration, but expiration was labored and sometimes required 4–5 seconds to complete. Upon examination, he demonstrated a barrel chest, a slight swelling of the extremities and a bluish color of the skin. He remarked that he had been losing weight over the previous few months. He wheezed almost continuously and demonstrated an intolerance to cold. He had unsuccessfully tried to give up smoking, and he now experienced severe coughing fits with almost every cigarette he smoked. Almost any stress appeared to heighten his symptoms, and exertion of any kind or degree caused exhaustion and occasionally prostration. His physicians were unable to treat his condition medically or surgically. His physical condition deteriorated to the point where he became an invalid, and he died at age 57.

Questions

From what disease was this man suffering?
What appears to be the cause of this disorder?
Describe briefly the histology of the lung.

Discuss the muscles of respiration and which ones might come into play in forced inspiration and expiration.

Discussion

This patient was suffering from *emphysema,* which basically involves a vacuolation of lung tissue combined with a loss of the normal elasticity of the lung. The overabundance of air in the lungs was responsible for the barrel shape of the chest, and the loss of elasticity caused the labored expiration and wheezing. Destruction of lung capillaries resulted in reduced oxygenation of blood and hence the bluish color of the patient's skin. Smoking further reduces the capacity for oxygen uptake as well as producing local irritation. Although irritants such as tobacco smoke or industrial pollutants appear to be contributing factors to emphysema, the exact etiology of this disease is still unknown.

The segmental bronchi (see Case 3–1) divide and redivide within the lobules (the microanatomical subdivisions of the lobes) of the lung, eventually subdividing into bronchioles. The bronchioles measure about 0.5–1 mm in diameter. The smallest of these are known as *terminal bronchioles.* From each of these structures arise approximately two short *respiratory bronchioles,* and these in turn give rise to a number of *alveolar ducts.* Each *primary lobule* of the lung consists of an alveolar duct; *alveolar sacs* and *alveoli* that arise from it; and associated nerves, blood vessels and lymphatics. The barrier across which gases enter and leave the blood stream consists of *alveolar epithelium, capillary endothelium* and the basement membrane(s) between the two.

The most important muscle of ordinary respiration is, of course, the *diaphragm.* During quiet inspiration the diaphragm descends and increases the superoinferior dimension of the thorax, causing air to enter the lungs. The *intercostal muscles* also contract, thereby elevating the ribs (sometimes referred to as the "pump-handle" mechanism) and thus increasing the anteroposterior dimension of the thorax. During deep inspiration the *accessory muscles* of respiration come into play. These are *scalenes* and *sternomastoids,* which raise

THE APPARITION OF THESE FACES

Zella Christensen

THANKS TO THE GHOSTS IN HIS HOUSE, MOREY LEFT FOR work an hour earlier than usual. He had fled to the metro to escape the dead, because all levelheaded people know that supernatural beings only appear when there is no one else around to witness them. On the orange line from Vienna into D.C., he kept looking over his shoulder to make sure his dead mother had not followed him. She was the principal ghost. She had woken Morey up by shaking his shoulder gently and whispering, "Honey, do you know what day it is?"

For a moment, the question so consumed Morey that he didn't wonder who had asked it.

"My birthday," he said after a sleep-muddled pause.

Then he sat straight up and fumbled for his glasses, because Morey lived alone, and he was sure he recognized that voice. When he put his glasses on, there she was, standing over him and smiling tenderly. Morey almost passed out from fear. Instead, he dove past the first and second ribs and the sternum, and the *pectoralis major* and *minor* and *serratus anterior*, which also elevate the upper ribs and indirectly the *erectores spinae;* the latter straighten the back and thereby spread the ribs. Quiet expiration is effected simply by the normal elastic recoil of the lungs and to some extent by the costal cartilages. It is this loss of lung elasticity in emphysematous patients that is responsible for their labored expiration. Contraction of the abdominal muscles, which subsequently forces the diaphragm upward, often is necessary in order for these individuals to expel air from the lungs.

CASE 3–4: FETAL LUNG DISEASE

History

A premature baby (born during the sixth month of gestation) demonstrated extremely difficult breathing and, despite various emergency attempts to save its life (i.e., positive pressure breathing), died some 30 hours after birth. Prior to death the baby became progressively cyanotic, and radiologic examination demonstrated fine reticulogranular markings on the lungs. Postmortem examination revealed atelectasis (collapse of the lungs), extensive congestion and a hyalin-like membrane lining the alveoli, which were dilated. The surface tension of extracts made from the baby's lungs was as much as four times higher than normal.

Questions

What is the name of this disorder?

What is thought to be its cause, and how does it result in the observed symptoms?

Discuss the disease in terms of alveolar structure and composition.

Discussion

This condition, a major cause of death in prematurely born babies, is known as *respiratory distress syndrome* (RDS) or *hyaline membrane disease* named after the glassy layer found in the alveoli during autopsy. It also occurs in increased incidence in babies born by cesarean section and in those born to diabetic mothers. It was originally thought among other things to be caused by aspiration of amniotic fluid or transudation of protein across the pulmonary capillary endothelium. It is now strongly believed to be due to a defect in the production of *surfactant*, a lipoprotein film that reduces alveolar surface tension and aids the alveoli in the recoil expulsion of air after inspiration.

The *alveoli* (see Case 3–3) are composed of *squamous pulmonary epithelial cells* (also known as small alveolar cells or type I pneumocytes) and *alveolar* or *septal cells* (great alveolar cells or type II pneumocytes). The alveolar wall is separated from the capillary endothelium by a basement membrane (basal lamina). The epithelial cells are flattened elements, whereas the alveolar cells are rounded or cuboidal. The latter contain in their cytoplasm concentric lamellae known as *multilamellar bodies* or *cytosomes*. These bodies appear to be rich in phospholipids, and the cells appear to secrete their contents into the alveoli. Considerable evidence is beginning to suggest that through this mechanism the alveolar cells produce and secrete the surfactant that is essential to normal lung action. This cell type appears to develop around the end of the sixth month of gestation.

Chapter 4

CARDIOVASCULAR SYSTEM

CASE 4–1: COLLATERAL ARTERIAL CIRCULATION

History

A 20-year-old male who had had a history of high blood pressure visited his physician with the complaint of noticeable subcutaneous "pulsations" occurring bilaterally approximately 2 cm on either side of the sternum. During the examination the physician also noticed similar pulsations around the borders of both scapulae. As a result of blood pressure measurements in both upper and lower limbs, it was determined that the patient was hypertensive in the head, neck and both upper extremities, but hypotensive in both lower extremities. A history revealed that he had frequent throbbing headaches. Radiographic examination of the thorax revealed that the inferior borders of the ribs, especially the posterior portions, were notched to a much greater degree than normal. The patient was scheduled for thoracic surgery, which resulted in a correction of the blood pressure abnormalities and a reduction and eventual elimination of the pulsations in the thoracic and scapular areas.

Questions

What specific condition was this young man suffering from?

Where did it occur in this case?

How did you determine its site, i.e., why, based on the symptoms described, could it not have occurred more inferiorly, for example, in the abdomen?

Why were the ribs notched on their inferior borders?

What caused the pulsations around the margins of the scapulae and along the lateral margins of the sternum?

Review the arterial anastomoses in the thoracic and abdominal walls and in the scapular area.

Would the condition be likely to be congenital or acquired?

Discussion

The condition in this case is known as *coarctation* (an abnormal narrowing or constriction) *of the aorta.* This phenomenon may occur anywhere from the origin of the brachiocephalic (innominate) artery to the bifurcation of the abdominal aorta. The most frequent site of coarctation, however, is between the left subclavian artery and the attachment of the ligamentum arteriosum. This region of the aorta is commonly constricted in the fetus and is referred to as the *aortic isthmus;* a dilatation distal to the isthmus, when present, is called the *aortic spindle.* A coarctation at this site or at one proximal to it is known as the *preductile* type. This type may involve the left subclavian artery or, in rare cases, it may even involve the brachiocephalic artery. In such cases it is obvious that various portions of the head, neck or upper extremities (depending on the specific site of the coarctation) would experience the same hypotension as the majority of the trunk and lower extremities demonstrated in our patient. In other words, any vessel and its branches located *proximal* to the constriction would be hypertensive, whereas those located *distally* would be hypotensive. If the preductile type of coarctation involves the brachiocephalic artery, most of the potential pathways for collateral circulation around the constriction are excluded, and survival time for individuals (predominantly newborn infants) so afflicted is very short.

The *postductile* type of coarctation (i.e., one occurring distal to the ductus arteriosus) often is compatible with life because an adequate collateral circulation is allowed to develop between the branches of the arch of the aorta and those distal to the constriction (see below). This is the type of coarctation that was present in this case. That the constriction in our patient was located distal to the left subclavian artery, yet still in the superior mediastinum, is supported by two important observations: (1) the patient was hypertensive in *both* upper limbs and, therefore, the coarctation was located beyond the left subclavian artery; and (2) the notched ribs were observed on the radiograph of the thorax. These notches are caused by collateral arterial pathways being established between branches of the aorta proximal to the coarctation and those distal to the point of stricture, mainly the *posterior intercostal arteries.* It was the enlargement of these arteries that caused the abnormal degree of notching. If the coarctation were to occur in the abdomen, it is doubtful that notching of the ribs would be observed. Although many of the same collateral pathways would be used, other segmental branches in the abdomen would share the load, so to speak, and the hypertension would be more widely distributed.

The specific anastomoses that permit arterial blood to be diverted from the aorta above the coarctation to the lower portion of the body are as follows (Fig. 4–1): (1) Branches of the *axillary artery (thoracoacromial, lateral thoracic, subscapular)* and the *subclavian artery* (*suprascapular, transverse cervical* and the first two *posterior intercostal arteries* from the costocervical trunk) anastomose with posterior intercostals lower down. Many of these arteries engage in an anastomotic network around the borders of the scapula. The suprascapular artery ramifies around the superior margin, the deep branch of the transverse cervical artery along the medial margin and the

Fig. 4–1.—Collateral circulation around the scapula between branches of the *subclavian* and *axillary arteries.* (Redrawn from Crafts, R. C.: *A Textbook of Human Anatomy.* Copyright © 1966, The Ronald Press Co., New York.)

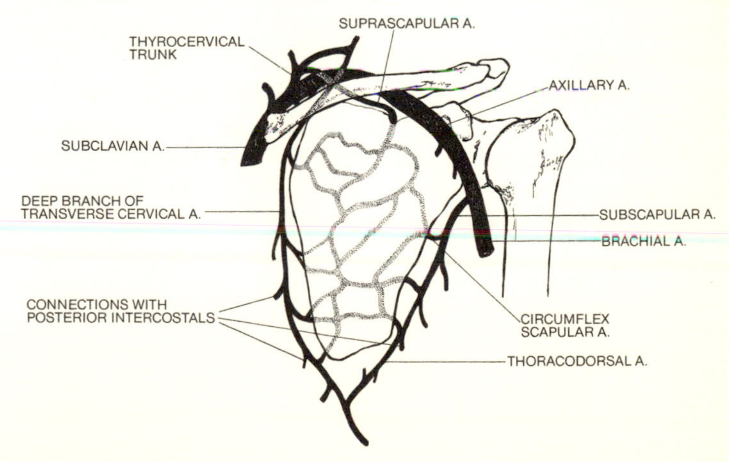

circumflex scapular and thoracodorsal arteries along the lateral border. These arteries anastomose with each other in the supraspinous and infraspinous fossae and, as mentioned above, with branches of posterior intercostal arteries in the dorsal body wall. These anastomotic channels also would come into play in the event of slowly formed occlusions of the distal portion of the subclavian artery or of the axillary artery (see Case 7–1). Enlargement of these arteries resulted in the pulsations around the margins of the patient's scapulae. (2) The *internal thoracic artery*, also a branch of the subclavian, gives off anterior intercostal branches that anastomose with posterior intercostal arteries in the upper six intercostal spaces. Below the sixth space the internal thoracic divides into the *superior epigastric* and *musculophrenic arteries*, which anastomose in the abdominal wall with the *inferior epigastric* and *deep circumflex iliac arteries*, respectively, thereby establishing a bypass between the subclavian and external iliac arteries. In addition, the musculophrenic gives off anterior intercostals to the lower six spaces, which anastomose with posterior intercostal arteries. The enlargement of the internal thoracic artery was responsible for the pulsations that the patient felt in the intercostal spaces just lateral to the sternum on each side. An additional anastomotic pathway might be present between the anterior spinal branch of the vertebral artery and various segmental arteries inferior to the coarctation.

Warwick and Williams[2] in the 35th British Edition of *Gray's Anatomy* state that coarctation of the aorta is congenital. On the other hand, Hollinshead[3] states, "Although some instances of this condition have been described as being congenital, it is not clear whether they really are or not." This area would therefore seem to be a potentially fruitful one for future research investigations.

CASE 4–2: ABDOMINAL VENOUS OBSTRUCTION

History

An elderly man presented with edema of the lower extremities and back and enlarged superficial veins of the lower limbs, abdominal wall and thoracic wall. Most prominent were the superficial epigastric and superficial circumflex iliac veins, the lateral thoracic veins and the thoracoepigastric veins, which connect the veins of the groin with those of the axillary area. There was no detectable accumulation of ascites fluid, and there did not seem to be any involvement (i.e., abnormal function) of the abdominal viscera. The man did complain of lower abdominal pain, however. A history revealed that his condition had been progressing slowly but consistently over a period of several months. Radiographs revealed a neoplasm in the abdominal cavity along the medial border of the right psoas major muscle at the level of the body of the third lumbar vertebra. A venogram revealed that the tumor was exerting pressure on the inferior vena cava to the point of almost totally occluding its lumen. The tumor was removed surgically, it was found to be benign upon pathologic examination and the patient made a normal recovery considering his age. Within a few weeks, the symptoms had disappeared.

Questions

How was venous blood from the lower half of the body able to bypass the obstruction in the inferior vena cava?

Would you expect that deep veins as well as superficial ones would be involved?

Review the tributaries of the inferior vena cava and the anastomotic connections that would allow the necessary collateral circulation in this case.

What specific characteristic of the development of this condition prevented a more serious or perhaps disastrous crisis to the patient?

Would the occlusion be more serious at the level of L. V. 1 or higher?

What additional problems would you expect to observe if this were the case?

Discussion

Like the aorta in the previous case, the inferior vena cava has numerous anastomotic channels that will ordinarily allow bypass of blood around an obstruction, provided the time course of the development of the obstruction is not too rapid. These connections occur both superficially (e.g., in the subcutaneous fascia) and deeply (e.g., in the retroperitoneal tissues of the posterior abdominal wall). The subcutaneous anastomoses are the most readily observable when engorged with blood, as evidenced by the prominent veins of the anterior abdominal and thoracic walls in this patient. These channels also may be used in hepatic portal vein hypertension, but the pattern is somewhat different, emanating in a spokelike fashion from around the umbilicus (see Case 5–1).

The tributaries of the inferior vena cava, from below upward, are the *common iliac, gonadal, renal, suprarenal, inferior phrenic* and *hepatic veins.* It should be noted that, while on the right side, these veins enter the inferior vena cava directly, and on the left the gonadal, suprarenal and sometimes the left inferior phrenic veins enter the left renal vein (Fig. 4–2). In addition to these tributaries, the inferior vena cava also receives the segmental *lumbar veins* on each side. The lumbar veins usually are connected by longitudinal channels, the *ascending lumbar veins,* on each side. Each of these is joined as it passes behind the diaphragm by the corresponding subcostal vein, forming the *azygos vein* on the right side and the *hemiazygos vein* on the left. The lumbar veins also possess numerous communications with the *vertebral plexus of veins,* which drains the back, the vertebrae and the structures that occupy the vertebral canal (see below).

The common iliac veins are formed by the union of the *internal* and *external iliac veins.* The internal iliacs drain the pelvis and most of the perineum, whereas the external iliacs, by way of the femoral and saphenous veins, drain the lower extremities, a small part of the perineum and the lower portion of the anterior abdominal wall. The internal iliac veins receive the *middle rectal (hemorrhoidal) veins* and indirectly the blood from the *inferior rectal veins* via the *internal pudendal vein.* Obstruction of the inferior vena cava, causing venous hypertension in the lower portion of the body, could result in retrograde blood flow in many or all of these veins. It will be recalled that the superior rectal vein anastomoses in the wall of the rectum with the middle rectal veins, and these in turn establish connections with the inferior rectal veins. The superior rectal becomes the *inferior mesenteric vein* in the abdomen and is one of the tributaries of the hepatic portal vein.

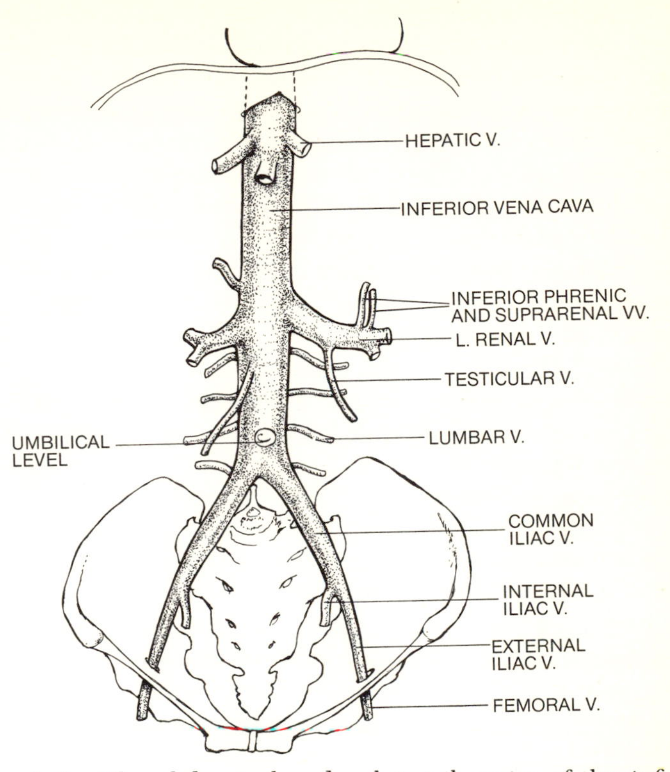

Fig. 4–2.—The abdominal and pelvic tributaries of the *inferior vena cava.* (Based on Becker, R. F., *et al.*: *The Anatomical Basis of Medical Practice* [Baltimore: Williams & Wilkins Co., 1971].)

Therefore, it might be expected that blood from the rectum and anal canal, which would ordinarily be destined for the internal iliac veins, could bypass the obstruction in the inferior vena cava via the following course: superior rectal vein→→ inferior mesenteric vein→→ hepatic portal vein→→ liver→→ hepatic veins→→ inferior vena cava→→ right atrium. The opposite is true in hepatic portal hypertension, in which hemorrhoids (enlarged middle and inferior rectal veins) often are a diagnostic symptom (see Case 5–1).

A second, more obvious, form of inferior vena cava bypass occurs between the veins that drain the lower abdominal wall

and those that drain the thoracic wall. The former may be divided into a superficial group *(superficial epigastric* and *superficial circumflex iliac veins),* which enter either the femoral or great saphenous veins, and a deep group *(inferior epigastric* and *deep circumflex iliac veins),* which drain to the external iliac vein. The superficial veins communicate via a series of interconnected longitudinal channels with the veins of the axilla (primarily the *lateral thoracic vein*). If a single, well-defined connecting vein is present, it is referred to as the *thoracoepigastric vein* (this vein, normally not apparent in human beings, is an obvious characteristic of the flank of the horse). The lateral thoracic vein drains to the axillary vein, which becomes the subclavian vein at the lateral border of the first rib; the subclavian veins drain to the *superior vena cava.* The inferior epigastric vein communicates with the *superior epigastric vein* within the substance of the rectus abdominis muscle; the veins accompany the arteries *(venae comitantes)* and, therefore, for the most part occupy a position between the posterior lamella of the rectus sheath and the rectus abdominis muscle. The superior epigastric vein is a tributary of the venae comitantes of the *internal thoracic vein;* the latter drains into the subclavian or brachiocephalic vein. Thus, venous blood may bypass the obstruction in the inferior vena cava and enter the superior vena cava by these deeper routes also.

The third way that blood may bypass the obstruction is via the segmental lumbar veins. These veins have connections with the *azygous system of veins* (see above) and hence with the superior vena cava, with the small but plentiful *retroperitoneal veins,* which carry venous blood from the walls of retroperitoneal organs to the hepatic portal system, and with the vertebral plexus of veins. The latter are valveless structures that form a more or less continuous plexiform connection with the pelvis and the cranial cavity as well as with the systemic veins in between. They have been widely implicated in the spread of malignant cells or disease processes from one part of the body to another (e.g., in the metastasis of prostatic cancer to the brain). Increased pressure in the inferior vena cava would therefore force blood in a retrograde direction in the lumbar veins and into the vertebral plexus. In normal inspiration, as intrathoracic pressure is decreased and intra-abdominal pressure increased, blood enters the vertebral plexus in the abdomen, ascends in the vertebral canal and then enters the systemic veins in the posterior thoracic wall. This blood may then enter the tributaries of the azygous system of veins (with which the vertebral plexus also is connected) to again eventually enter the superior vena cava. The opposite flow occurs in normal expiration, although such flow would be reduced or eliminated in cases of inferior vena cava obstruction.

In very early development in the human being as well as in other mammals, numerous endothelial tubes invade the mesenchyme and, as the various organs develop, some of these are chosen and others are not. The reasons for this, whether genetic or due to changing patterns of blood flow or blood pressure, are unknown. Most of the unchosen channels degenerate, but those that do not may form anastomotic channels if a need for such develops (e.g., the thoracoepigastric vein in the event of obstruction of the inferior vena cava or in hepatic portal hypertension). The significance of such "residual" channels lies in the fact that they will develop according to need, provided this occurs *slowly*, as was the situation in this patient (see below). In addition, these channels may form supernumerary or anomalous vessels and undoubtedly furnish a significant opportunity for the normal vascular variation observed in all human beings.

If the obstruction is above the level of the kidneys, the condition is much more serious, and the symptoms are more striking. According to Durham,[4] involvement of the hepatic veins usually is associated with problems in the function of the liver, kidneys and additional abdominal organs. This usually is accompanied by jaundice, ascites, vomiting and diarrhea, and albuminuria also may be evident. The prognosis in inferior vena cava obstruction is dependent on the site and on causative factors. For example, the condition is compatible with life in numerous cases, but death almost always results from rapid occlusion superior to the level of the renal veins.

CASE 4–3: HEART PAIN AND MYOCARDIAL INFARCTION

History

A 46-year-old male executive was engaged in one of his thrice-weekly tennis matches when he first noticed a diffuse substernal pain that radiated across his chest to his left shoulder and part way down the left arm. Having been an athlete almost all his life, however, he continued his tennis playing as well as swimming and occasionally playing handball. The bouts of pain became recurrent with each strenuous exercise and subsided during rest. Thinking the pain would go away and not wanting to endanger his masculine image, he did not seek medical assistance. During one particularly long and strenuous game of handball, he experienced a crushing pain in the chest, was overcome by fear of impending disaster and subsequently slumped to the floor. He was rushed to the nearest hospital and placed immediately in the intensive care unit. Tests revealed, among other things, an elevated serum cholesterol level and an abnormal ECG. The patient was placed on anticoagulant and vasodilatory drug therapy and prescribed total bed rest for 10 days, followed by a long period of convalescence that included improved diet, rest, moderate but regular exercise, abstinence from tobacco and reduction of stress. His condition improved steadily and he has suffered no recurrences of the original disorder.

Questions

What is the term for the recurring pain experienced by the patient upon exertion?

What is the cause of this condition?

Very briefly describe the nerve supply of the heart.

Why was the pain perceived in the left shoulder and arm as well as in the thorax?

Describe the pattern of arterial blood supply to the heart.

How do you account for the fact that the man survived this ordeal when so many other instances of the same problem are fatal?

Discussion

This patient was suffering from *angina pectoris,* pain caused by a reduction in the normal volume of blood flow to the heart. The underlying cause usually is a narrowing of the diameter of one or more of the major arteries of the heart as the result of atherosclerotic plaque formation on the internal walls of these vessels. The patient's elevated cholesterol level combined with the pain upon exertion would tend to support a diagnosis of reduced coronary artery capacity. When the patient was at rest, the arteries apparently were capable of carrying an adequate blood supply, but with increased demand, as occurs in exercise, the heart became deprived of its required oxygen level, resulting in ischemia and pain. This condition finally resulted in a "heart attack" *(myocardial infarction),* in which a portion of the heart was sufficiently deprived of oxygen for a long enough period to result in actual death of some of the tissue. The abnormal ECG was evidence of an altered nerve conduction pattern in the heart, strongly suggestive of damage to a portion of the heart muscle, with eventual scar tissue formation.

Pain of myocardial infarction often is perceived over the left shoulder and arm as an example of what is known neurologically as *referred pain.* The heart is supplied by the autonomic nervous system through parasympathetic fibers from the vagus nerve (cranial nerve X) and through sympathetic fibers from the upper four segments of the thoracic portion of the spinal cord ($T_{1\text{-}4}$). The two autonomic subdivisions are antagonistic in their action on the heart, the sympathetic causing the heart rate to increase, whereas the parasympathetic is responsible for reducing the rate. The fibers from both subdivisions intermingle and ramify around the aorta as the *cardiac plexuses of nerves.* In addition, sensory fibers travel from the heart and enter the dorsal roots of the upper four or five thoracic spinal nerves. These same nerves receive afferent fibers from the shoulder and arm (mainly T_1 and T_2) and from the thoracic wall. Although the mechanism of referred pain is still poorly understood, somehow pain impulses from the heart become intermingled with afferent fibers from the periphery of the body in such a way that somatic pain also is felt (in this case from the left shoulder and arm). This phenomenon is common in other parts of the body as well, e.g., pain from the stomach or from the diaphragm is frequently referred to the patient's back.

The pericardium receives arterial blood from the *pericardiacophrenic branches* of the *internal thoracic arteries* and also small twigs from adjacent blood vessels such as *bronchial, esophageal* and *superior phrenic arteries*. The heart proper is supplied by the *right* and *left coronary arteries* (the only branches of the ascending aorta) originating from the right and left aortic sinuses, respectively. The right and left sinuses correspond to the right and left cusps of the aortic semilunar valves. Each coronary artery appears on the corresponding side of the pulmonary trunk, between it and the right or left auricle (Fig. 4–3, A). The right artery courses to the right and inferiorly in the *coronary (atrioventricular) sulcus*. It gives branches to the right atrium and ventricle and a fairly constant *right marginal branch* and then courses around to the posterior surface of the heart. The left coronary artery gives off the *anterior interventricular (descending) artery*, which descends toward the apex of the heart in the slight groove marking the anterior extent of the interventricular septum and then continues to the left in the coronary sulcus as the *circumflex artery*. The latter supplies the left atrium and ventricle, giving rise to the fairly constant *left marginal artery* and then continues around to the back of the heart. On the posterior surface of the heart (Fig. 4–3, B), the right coronary gives off the *posterior interventricular (descending) branch*, which descends in the groove marking the posterior limit of the interventricular septum. It then continues across the midline to anastomose with the circumflex artery usually deep within the myocardium. This is one of the important communications that allows collateral circulation between the right and left coronary arteries. The other occurs near the apex of the heart between the anterior and posterior interventricular arteries. Thus, provided these anastomotic channels are adequate, an occlusion in one or the other coronary might be tolerated due to retrograde flow of blood through one or more of these collateral pathways to the area of the clot. In addition, there are numerous *extracardiac anastomoses* between the coronary arteries and those that supply the pericardium (see above) as well as other extrinsic branches to the heart (posterior intercostal, tracheal

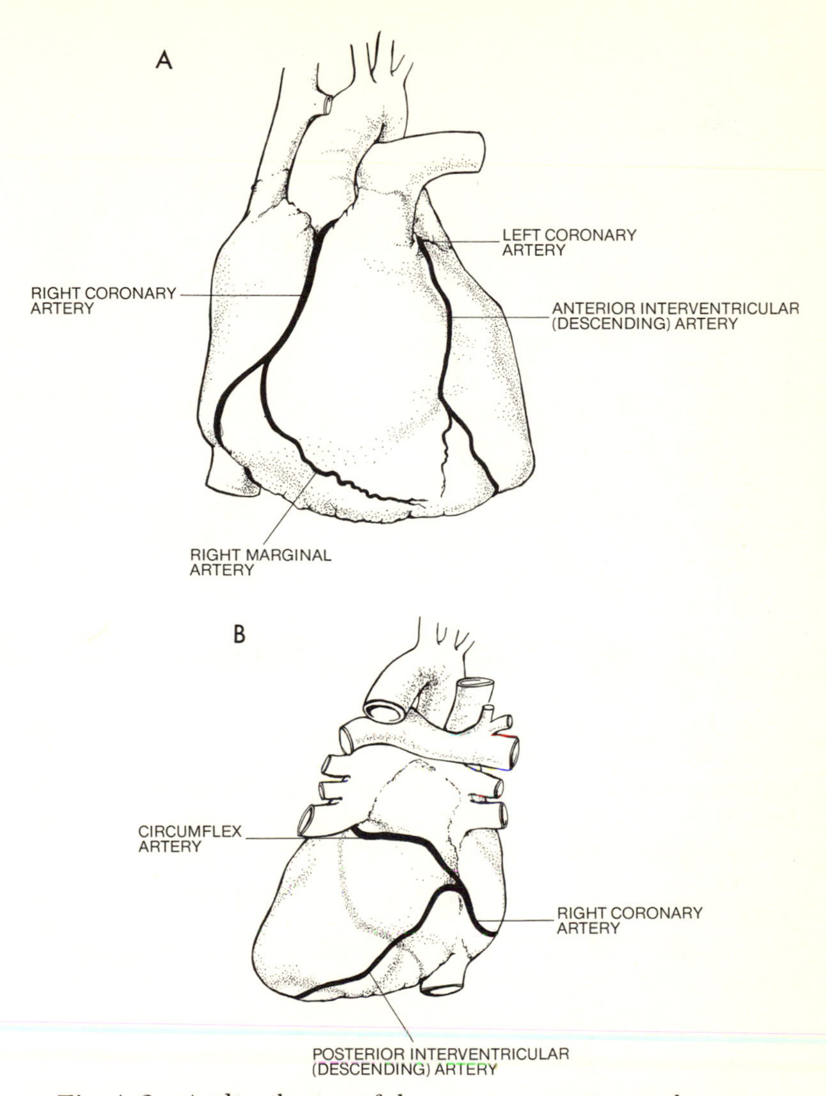

Fig. 4–3. – **A,** distribution of the coronary arteries on the anterior surface of the heart. **B,** distribution of the coronary arteries on the posterior surface of the heart. (Based on Gardner E., *et al.: Anatomy* [3d ed.; Philadelphia: W. B. Saunders Co., 1969].)

and in some cases small branches emanating directly from the descending thoracic aorta).

These anastomoses usually occur on a capillary or precapillary basis, although small arteries from different sources (i.e., small branches of the two coronary arteries) sometimes may have direct connections. Ordinarily, not all of these channels are in use but may be called into action under the conditions of a slowly developing occlusion. With regard to the patient in this case, two factors were working in his behalf. One is that he had been athletically active most of his life and, by regularly exercising his heart muscle, maintained a relatively profuse arterial circulation to the organ. The other is that the coronary arterial occlusion that eventually manifested itself as a myocardial infarction developed gradually, as evidenced by the slow onset of the angina pectoris. This allowed time for additional collateral pathways to develop. A sudden coronary occlusion often results in death of too large a segment of the heart to be compatible with life.

CASE 4–4: HEART CONSCIOUSNESS

History

A man, age 36, was driven by his wife to his physician's office, complaining primarily of severe pain in the left side of the chest. He had been cutting the lawn with a manual mower when the pain began. It was preceded by breathlessness, heart palpitations, dizziness and profuse sweating. It was a relatively hot day, and he had felt fatigued after cutting the first couple of swaths of grass. He had had these symptoms off and on for the previous 4 or 5 years, also complaining variously of nervousness, headaches, trembling, flushing, muscle cramps, insomnia and excessively dry mouth. A preliminary investigation by the physician revealed slightly elevated blood pressure (145/85), a pulse rate of 96 and a slightly elevated temperature (99.4F). He was unable to identify any specific organic disease, however, and referred the patient to a cardiologist in order to exclude all possibilities. The ECG was normal, as were the cardiac shadow and heart size in radiograms. Stethoscopy revealed no murmurs either at rest or following physical exertion. The serum cholesterol level was normal, and there was no evidence of coronary artery disease. During the course of the investigation, as the patient relaxed and the chest pains subsided, the blood pressure and pulse rate returned to normal.

The temperature subsided to normal also. The man was told by the cardiologist that nothing was wrong with his heart and was sent home with instructions to "take it easy for a while." He was prescribed a mild tranquilizer. The bouts of chest pain and related symptoms recurred, however, particularly after strenuous exercise or during anxiety-laden situations or those in which he became angry. He went from physician to physician insisting that he had "heart trouble" and fearing that he was going to drop dead any minute if something was not done to help him. Finally, one understanding doctor, recognizing the emotional aspects of the man's complaints, referred him for psychotherapy.

At first, the man rejected the idea that his disorder might have any psychologic implications, but he soon began freely to pour out his story. He had been a delicate, sensitive, nervous child whose father complained incessantly about his weak heart. The boy was able to perform all the physical acts of his peers, however, and he demonstrated no deficiency of normal stamina. However, at age 10, his father, then 31, died of a heart attack after a day's hard physical labor. The boy went to live with his uncle (his father's brother), who also complained of heart trouble and died 3 years later. Shortly thereafter, the boy began having palpitations and, fearing that he would die as his father and uncle had, especially if forced to perform physical work, he began to avoid almost all forms of physical activity. At age 16 a bout of strep throat prompted a visit to a doctor who incorrectly informed him that his heart was damaged as a result of the disease. During his Army induction physical he was told, also incorrectly, that his heart was enlarged. He subsequently became a semi-invalid, moving around very slowly, sighing a great deal and becoming easily fatigued. As time went by, however, the heart attack he so much expected never materialized. He began to increase his physical activity until the bouts of chest pain in his 30s convinced him he had been right all along. With the aid of the psychiatrist and the insistence of specialists that his heart was perfectly normal, he recovered to the point where he could lead a normal life. He never completely lost his inordinate fear of strenuous physical exercise, however.

Questions

What is the name of this disorder?

What do you think might have been some of the suspected causes prior to the conclusion that it probably resulted purely from emotional factors?

What is meant by psychosomatic medicine? Is there any rational basis for this aspect of medicine?

Describe briefly the structures that contribute to the cardiac shadow in x-rays.

Where are the various valve sounds of the heart best perceived stethoscopically on the thoracic wall?

How can differences in body type, changes in body position or differences in the respiration phase alter the "normal" shape and/or position of the heart?

Discussion

This condition is known as the *Da Costa syndrome*, named after the individual who described it in 1871 as a result of observations made on soldiers during the Civil War. It is also known as *neurocirculatory asthenia, cardiac neurosis, effort syndrome, soldier's heart, autonomic imbalance, irritable heart* and *disordered heart action*. It is a long-recognized disorder arising out of stressful situations, especially those in which fear is a significant element, i.e., during wartime (see case 8–1, which discusses the body's general reaction to frightening situations). An estimated 60,000 British soldiers suffered from this syndrome in World War I. It also shows up in young men and women who have undergone disturbing disruptions in their emotional lives. According to Wood,[5] it has been diagnosed as rheumatic carditis, toxic myocarditis, weak or tired heart, cardiac strain, anemia, thyrotoxicosis, pulmonary tuberculosis, pleurisy, influenza, angina pectoris or malingering. He noted that four theories have existed to attempt to explain the left precordial pain, that it is (1) cardiac; (2) referred pain from "faulty function of muscles or ligaments about the spine"; (3) myalgia (muscle pain) resulting from fatigue of the accessory muscles of respiration or (4) imaginary. Wood strongly doubted that it was imaginary primarily because of the uniformity of its character and behavior and because "medical colleagues who have suffered from it are good witnesses of its reality." No consistent organic disorder has ever been identified with or ascribed to this syndrome, however, and it is thought today to be entirely psychogenic. Perhaps the best evidence for the latter conclusion is that the symptoms usually abate or disappear altogether upon removal of the fear-causing stimulus (e.g., being sent home from the war). Although there seems to be a hereditary predisposition to the Da Costa syndrome, according to Durham,[4] the physiologic disturbances usually can be attributed to a large number of distressing psychogenic problems that become evident during psychiatric investigation.

Psychosomatic medicine deals with the relationships between the emotions and disease. The word is derived from the Greek words *psyche* meaning "soul" and *soma* meaning "body." The term originated in Germany in the 1920s but was first used in the United States by Dunbar[6] in the early 1930s. The concept that emotional processes can cause physical diseases can be traced back in the history of medicine approximately 4,500 years. Nearly all physicians recognize the importance of the emotional component in such disorders as peptic ulcer, hypertension, asthma, allergies and many others. Recently, the relationship of the psyche to many instances of impotence, to physical paralyses, to blindness and even to certain malignant growths has become apparent. For example, a relatively frequent observation in certain cancer patients is that they have undergone severe emotional shocks, e.g., loss of a loved one, just prior to the onset of the disease.[7, 8] The disease in these patients was linked to a feeling of hopelessness. Many physicians are reluctant to accept the probable extent of the psychosomatic origin of physical disease. Nevertheless, a Johns Hopkins Hospital survey revealed that approximately three of four patients among the study cases presented with diseases that were psychogenic. A leader in bringing the awareness of psychosomatic principles in medicine to the American public was the physician-author Frank G. Slaughter[9] in his book *Medicine for Moderns* (later published under the title *Your Body and Your Mind*).

The *cardiac (cardiovascular) shadow* (Fig. 4–4) is formed on the right side from superior to inferior by the right brachiocephalic vein, superior vena cava, right atrium and occasionally a small segment of the inferior vena cava. It should be noted that the right ventricle does not contribute to the right

"Now you're worth more dead than alive."

I am worth more dead than alive. No doubt about it.

I open the policy and check the language again, just like I do every day. The suicide clause expired two weeks ago. Fifty-thousand dollars. That would take care of my kids for years. All I have to do is take some pills, and they'll never have to eat macaroni again.

I put the policy back in its beat up envelope. Today, I decide to keep going.

I'll take one of the toilet paper rolls from the company bathroom. No one will know. I've got to have toilet paper.

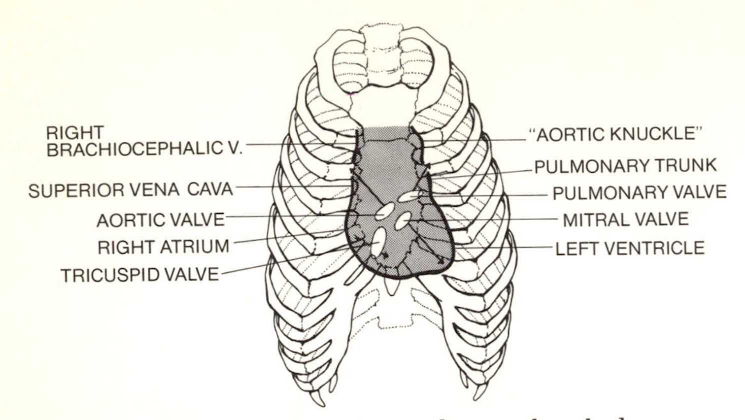

Fig. 4–4.—Relationship of the cardiovascular shadow seen on radiographs of the thoracic cage. The transmission of valve sounds to their points of clearest audibility are indicated by the *arrows*.

margin of the cardiac shadow. The left margin of the shadow is formed from above to below by the arch of the aorta (radiographically the so-called "aortic knuckle"), the pulmonary trunk, the left auricle, the left ventricle and perhaps some extrapericardial fat.

The right border of the heart in its so-called "normal" position lies about 1 cm to the right of the right margin of the body of the sternum. The junction of the heart with the great vessels entering and leaving it superiorly lies at about the *sternal angle* (the junction of the manubrium with the body of the sternum; this junction is marked by a readily palpable transverse bony projection, which in turn marks the level of the second costal cartilages and is important in identifying intercostal spaces, the second lying immediately inferior to the second costal cartilage and rib). The left border of the heart diverges to the left from superior to inferior so that the *apex* of the heart lies in the left fifth intercostal space, some 6–9 cm lateral to the midline. The inferior *(diaphragmatic)* surface of the heart lies at the level of the *xiphisternal joint*. It should be noted that, although the above description is probably the most frequently observed, considerable variation occurs in this pattern so that, for example, the lower borders of normal hearts have been found as far as 4.5–5 cm below the level of the xiphisternal joint.

The *aortic semilunar valve* transmits its sound to the right second intercostal space, whereas that of the *pulmonary semilunar valve* is best heard at the left second intercostal space. The *tricuspid (right atrioventricular) valve* is heard behind the inferior aspect of the sternum in about the midline. The *bicuspid (left atrioventricular* or *mitral) valve* transmits its sound to the apical area of the heart (fifth intercostal space, approximately 7 cm from the midline), but it may be found outside the actual cardiac area.

The position and shape of the heart depend on body shape, size, physiologic status and body position. Normal hearts usually are spoken of as being *oblique* in position. However, in short, stocky individuals, during pregnancy or in young children, the heart assumes a more *transverse* position (i.e., higher and flatter). In tall, thin people the heart is more *vertical* (longer and thinner, with a greater inferior extent). In the upright position the axis of the heart (i.e., from base to apex) is directed from right to left and somewhat anteriorly. Lying on the right side tends to straighten the heart out in the median plane of the body, whereas lying on the left side accentuates the left-sided orientation of the heart. Indeed, the apex may be found quite close to the left side of the thoracic wall, and many normal individuals can feel their heart beat while lying on the left side. The normal position of the heart is described radiologically at maximum inspiration ("Take a deep breath and hold."). The fibrous pericardium is anchored securely to the central tendon of the diaphragm and, as a result, the heart moves up and down with movement of the diaphragm. During expiration, as the diaphragm ascends, the heart may change position up to one vertebral level. Similarly, in recumbency, the abdominal organs tend to shift cranially, resulting in a higher than normal position of the diaphragm and heart. Approximately the same magnitude of shift occurs in death and should be taken into account when analyzing the position of the heart in the cadaver.

CASE 4–5: STAB WOUND IN THE THORAX

History

A 26-year-old cowboy was stabbed in the left side of the chest with an icepick during an altercation in a bar. The wound was relatively shallow (it penetrated to a depth of less than 2 cm) and, other than the fact that it resulted in considerable local pain and some bleeding, it did not at the time cause the young man a great deal of concern. Within a few days, however, he began to complain of weakness, pain in the chest, dyspnea and swelling of the lower extremities, particularly around the ankles. A medical examination revealed blood in the pericardial cavity accompanied by disordered and weak action of the heart. He was placed immediately in the hospital, treated surgically, kept under observation for several days and finally released. Follow-up examinations revealed no recurrence of the problem.

Questions

What is the term used for compression of the heart caused by accumulation of fluid within the pericardial cavity?

Describe the pericardium from a gross anatomic, histologic and embryologic point of view.

Why was this injury not fatal to the cowboy? Under what conditions might it have been?

What do you suppose was the treatment of choice in this case?

Discussion

The clinical condition demonstrated by this young man is *tamponade,* a compression of the heart resulting from effusion of blood or other fluid into the pericardial cavity. It can be caused by rupture of the heart, as occurred in this case. If the rate of flow of blood into the pericardial cavity is very rapid and if escape from the cavity is negligible or nonexistent, heart compression becomes acute, venous return is impeded and death ensues rapidly. If, however, as occurred in this case, accumulation of blood is slow (apparently, the ice pick simply nicked the surface of the heart), symptoms are slower to appear. Nevertheless, a few cubic centimeters of fluid above that normally found in the cavity (see below) are sufficient to cause serious symptoms, as occurred in this case, and lead to eventual death if not removed. The reason for this lies in the unyielding nature of the pericardium. Very slow accumulation of fluid, however, may result in a distention of the pericardium attended by only moderate symptomatology or none at all.

The *pericardium* (Fig. 4–5) is a serous sac with parietal and visceral layers, separated by the *pericardial cavity*, a potential space that is moistened by a film of fluid. The visceral layer of the serous pericardium is the same as the *epicardium* that covers the external surface of the heart. When this terminology is used, only the parietal layer with its fibrous external covering is referred to as the pericardium proper. Even though the fluid is so arranged in the pericardial cavity that it amounts to no more than a film in most places, the total volume in health may be as much as 50 ml.[10] The *serous pericardium* consists of a mesothelium backed by a layer of connective tissue. The mesothelium is a simple squamous cell layer with interlocking serrated edges. The watery *serous exudate*

Fig. 4–5. – The margins and posterior attachments of the pericardium after removal of the anterior portion of the pericardium and the heart. Note the *transverse* and *oblique sinuses.*

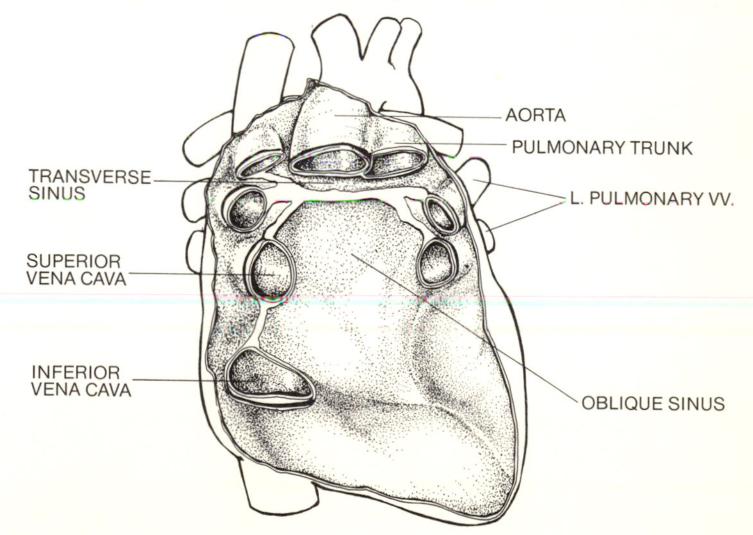

within the pericardial cavity allows the heart to move freely during systole and diastole. In certain pathologic states, adhesions may occur between the parietal and visceral layers of the serous pericardium that tend to limit the normal action of the heart. The *fibrous pericardium,* which lies external to and is adherent to the parietal layer of the serous pericardium, is a dense sheet of collagenous fibers with some elastic fibers in its deeper portions. It blends with the adventitia of the great vessels entering and leaving the heart, whereas the serous pericardium is reflected from the visceral to the parietal layer along the connections of these vessels with the heart. The fibrous layer is bound firmly to the diaphragm via the *pericardiacophrenic ligament* and to the sternum via the *sternopericardial ligaments.*

The heart develops embryologically as a simple tube with the arterial end directed cranially and the venous end caudally. As development proceeds, the heart bends on itself so that the venous end comes to lie behind the arterial end. As a result, two pericardial sinuses are formed, the first of which lies between the opposing layers of pericardium that coat the arterial and venous portions of the heart. This is the *transverse sinus of the pericardium,* which can be demonstrated by passing a finger laterally between the aorta and pulmonary trunk anteriorly and the atria and superior vena cava posteriorly. The other is the *oblique sinus of the pericardium,* which is the line of reflection of the serous pericardium at the base of the heart where the pulmonary veins enter the left atrium. If the pericardial cavity is opened from the front and one or more fingers are inserted beneath the apex of the heart and extended superiorly, they will be stopped by the reflections of pericardium around these veins. This is the oblique sinus.

The heart within its pericardial covering occupies the entire middle mediastinal portion of the thoracic cavity. The pericardium is supplied by the internal thoracic arteries via their *pericardiacrophrenic* branches. Some arterial blood also is supplied by pericardial branches of bronchial, esophageal and superior phrenic arteries (see Case 4–3 for a discussion of how these arteries may participate in extracardiac anastomosis). The pericardium is supplied by the *phrenic nerves,* which descend along its lateral surfaces en route to the diaphragm. The epicardium is insensitive to pain.

The pericardium may be reached and drained, as was done in this case, by excision of the left fifth costal cartilage, ligature and division of the internal thoracic artery and detachment of the transversus thoracis muscle from the sternum. In our patient, bleeding from the surface of the heart had ceased by the time drainage was performed, and no further problems were encountered by the patient.

CASE 4–6: TETRALOGY OF FALLOT

History

A newborn baby demonstrated marked cyanosis and, after lengthy examination and numerous tests, was determined by a team of physicians to be suffering from the *tetralogy of Fallot.* An operation was performed that allowed the infant to develop relatively normally.

Questions

Describe this condition and some of its variations with regard to its anatomic manifestations and the developmental defects that bring it about.

What type of operation could be performed that would at least partially alleviate the cyanosis?

Discussion

The tetralogy of Fallot (Fig. 4–6) consists of (1) pulmonary stenosis, (2) dextroposition ("overriding") of the aorta, (3) interventricular septal defect and (4) hypertrophy of the right ventricle. The cyanosis occurs as a result of the overriding of the aorta across the interventricular septal defect into the right ventricular area. As a result, some venous blood from the right ventricle as well as the arterial blood from the left ventricle is ejected into the aorta. This situation also results in increased pressure in the right ventricle and, consequently, the enlargement of the right ventricular wall. In addition, due to the stenosis of the pulmonary artery, the lungs do not receive their

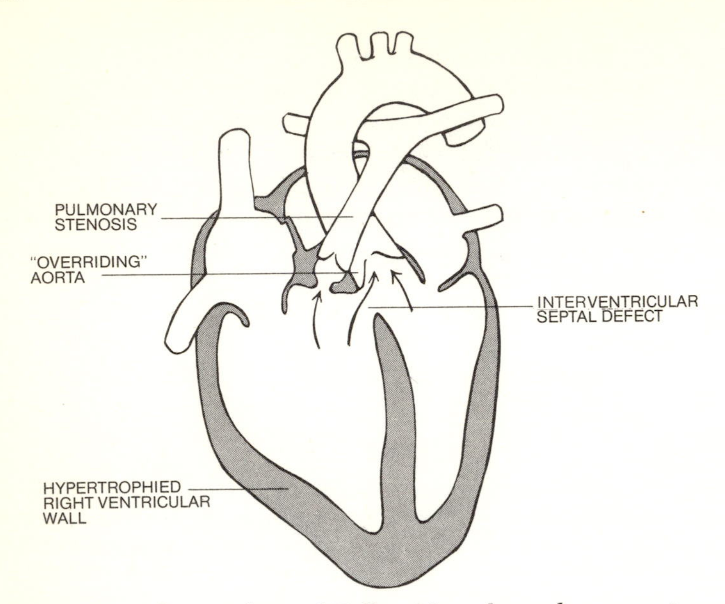

Fig. 4–6.—The tetralogy of Fallot. Note the *pulmonary stenosis, "overriding aorta," interventricular septal defect* and *hypertrophy of the right ventricular wall* (redrawn from Langman[11]).

normal proportion of blood, and oxygenation is impeded. Some texts describe a *trilogy of Fallot,* which consists of the pulmonary stenosis, but an interatrial rather than an interventricular defect. In *Fallot's pentalogy,* the four defects associated with the tetralogy are present plus an interatrial septal defect. Both the trilogy and the pentalogy are less common than the tetralogy. Some authors do not believe that the right ventricular hypertrophy should be included as part of the tetralogy per se since it is basically the result of the overriding aorta.

In the normal embryology of the heart, the *truncus arteriosus* (that portion of the *bulbus cordis* that develops into the proximal portions of the aorta and pulmonary artery) is partitioned into approximately equal halves by a septum. This occurs at about the 6-mm stage of development. The *conus cordis,* that portion of the bulbus just proximal to the truncus, also is divided during this time, the septum becoming the superiormost portion of the interventricular septum. In the tetralogy of Fallot, there is an unequal division of the truncus and conus caused by anterior displacement of the septum. This, in turn, results in the narrowed outflow channel from the right ventricle (pulmonary stenosis), the enlarged and overriding outflow channel from the left ventricle (the aorta) and the interventricular septal defect. In simple terms, the truncus and conus are unequally divided, resulting in a smaller than normal pulmonary artery and a larger than normal aorta (in their proximal aspects) plus the septal defect.

The operation, which is performed to partially correct this condition, is known as a *Blalock shunt* or the *Blalock-Taussig operation.* In this operation one of the subclavian arteries is surgically anastomosed to the pulmonary artery in order to shunt some of the systemic arterial blood to the pulmonary circulation and hence overcome some of the cyanosis.

REFERENCES

1. Crafts, R. C.: *A Textbook of Human Anatomy* (New York: The Ronald Press Co., 1966).
2. Warwick, W., and Williams, P. L.: *Gray's Anatomy* (35th British ed.; Philadelphia: W. B. Saunders Co., 1973).
3. Hollinshead, W. H.: *Textbook of Anatomy,* (2d ed.; Hagerstown, Md.: Harper & Row Inc., 1967).
4. Durham, R. H.: *Encyclopedia of Medical Syndromes* (New York: Paul B. Hoeber, Inc., 1960).
5. Wood, P.: Da Costa's syndrome (or effort syndrome), Br. Med. J. 1:767, 1941.
6. Dunbar, F.: *Emotions and Bodily Changes* (New York: Columbia University Press, 1954).
7. LeShan, L.: An emotional life-history pattern associated with neoplastic disease, Ann. N. Y. Acad. Sci. 125:780, 1966.
8. Schmale, A. H., Jr., and Iker, H. P.: The affect of hopelessness and the development of cancer, Psychosom. Med. 28:5, 1966.
9. Slaughter, F. G.: *Medicine for Moderns* (New York: Julian Messner, Inc., 1947).
10. Ham, A. W.: *Histology* (7th ed.; Philadelphia/Toronto: J. B. Lippincott Co., 1972).
11. Langman, J.: *Medical Embryology* (3d ed.; Baltimore: Williams & Wilkins Co., 1975).

Chapter 5

GASTROINTESTINAL SYSTEM

CASE 5–1: ALCOHOLIC CIRRHOSIS OF THE LIVER

History

A 63-year-old male with a long history of chronic alcoholism was admitted to the hospital in serious condition. He was bleeding from the mouth, was apparently suffering from malnutrition and frequently lapsed into a comatose state. Careful examination revealed that he also had hemorrhoids and blood in the stools, indicative of bleeding somewhere along the gastrointestinal tract. The anterior abdominal wall was distended due to an accumulation of ascites fluid within the peritoneal cavity. In addition, prominent bluish lines radiated subcutaneously from the umbilicus, and bilateral lines of this nature extended generally from the inguinal region to the axilla. The patient was placed on intravenous feeding and later, when he was able to take solid food, was given a high protein diet generously supplemented with vitamins and minerals. Bleeding was arrested by a balloon inserted into the esophagus and inflated; this served to compress the hemorrhaging blood vessel(s). The ascites fluid was withdrawn through a needle inserted into the peritoneal cavity. Liver tests and a needle biopsy of hepatic tissue revealed severely subnormal functions of this organ and histologic evidence of a ubiquitous replacement of normal tissue by fat and collagenous connective tissue. As a result, there was impedance of the normal flow of venous blood through the liver, leading to an increase in the normal blood pressure within the hepatic portal vein (portal hypertension). Therefore, once the patient's strength was restored to a reasonable level, he was scheduled for remedial surgery. The operation was successful, and shortly thereafter the patient's symptoms began to subside. He was released from the hospital 10 days later with the firm admonition by his physician to abstain from all alcoholic intake whatever while maintaining a properly balanced nutritional intake.

Questions

Review the normal flow of blood from the intestines through the liver to the heart.

Jerry is looking for a reason to fire me already. I've been late three times in the past two weeks. "We need you to be here at eight o'clock sharp to answer the phones," he says as he scolds me in his office. "Eight o'clock. When the phone rings at eight o'clock, I want to hear your cheery voice!"

And, of course, he calls every morning at eight o'clock.

I did okay at first, but then I started getting notices that there's no supervision at the school before 7:45am. They threatened to call Child Protective Services if the kids were brought to school too early again. So now I drop Matt and Bobby off a couple of blocks away and tell them to walk slowly and hide if they get there too soon. But it rained twice last week.

Last week was horrible. I had to leave Amanda home alone with the stomach flu. She was still vomiting when I left. She cried. Even at twelve years old. She's usually so grown up and a little bitchy, but she needed me then. "Mommy, please stay," she begged. But what could I do?

Now Amanda throws her spoon into her bowl.

"I hate this!" She whines. "Why'd you make daddy leave?"

There is a collective gasp. Matt looks down and quickly takes a bite. Bobby looks back and forth between his big sister and me. What do I say? Their father is a deadbeat who has abandoned us and run off with some bimbo who has time to go drinking and dancing at the bars. But I can't tell them that. At this moment, I hate my children almost as much as I hate their father.

I leave the table in silence.

"Why'd you say that?" I hear Matt whisper to Amanda.

"I hate them both!" she screams. Her chair screeches on the floor as she jumps out of it and runs to her room. She slams the door hard.

Bobby begins to cry. I hear Matt try to soothe him. "It'll be all right. Everything will be okay."

In my bedroom, I pick up the envelope again. Life insurance—I always make sure to pay this bill first. When we first bought the policy, Larry had made a joke about it.

What is meant by a "portal system?"

Identify the various anastomoses between the hepatic portal vein and tributaries of the superior and inferior venae cavae that would allow a considerable percentage of the patient's blood to bypass the obstruction in the liver.

Describe how each of the symptoms observed in this case was related to the collateral circulation (portal-caval or portal-systemic anastomoses).

Are there any connections that would not result in clinically recognizable symptoms in obstructive liver disease?

What type of operation do you think could be performed that would result in a reduction of the clinical symptoms observed in this case?

Discussion

The symptoms described in this case are the result of *alcoholic cirrhosis of the liver,* a disease that either directly or indirectly causes normal hepatic tissue to be replaced by connective tissue. These symptoms, which are mainly related to increased blood pressure in the hepatic portal vein, also are observed in carcinoma of the liver. In both cases the disease process tends to impede the normal flow of blood through the liver. This case provides an excellent example of the frequently intimate relationships between anatomical structure and associations and clinically observable disorders. The student is encouraged, in this case as in the others, to attempt to reason out the specific relationship between a given symptom (i.e., hemorrhoids) and the anatomy of the liver and its venous connections.

The *hepatic portal vein* is formed by the union of the *splenic vein* with the *superior mesenteric vein.* The *inferior mesenteric vein* also is a tributary of this system, either joining the splenic or superior mesenteric vein or the angle where these two veins meet. The portal vein occupies a position near the free edge of the lesser omentum (hepatoduodenal ligament), posterior to the common bile duct and the hepatic artery proper. It divides at the porta hepatis of the liver into right and left branches, which further divide to supply the *hepatic lob-*

ules. These are polygonal structures made up of cords of liver cells separated by sinusoids. At the center of each is a *central vein.* Venous blood carrying nutrients and toxic materials from the intestines enters the lobules from the peripherally located portal venules and percolates through the sinusoids where exchanges occur between it and the liver cells. The thus altered blood then enters the central vein. The central veins from the numerous lobules coalesce to eventually form the *right* and *left hepatic veins* (sometimes three or four), which enter the *inferior vena cava.* Blood from this large vein enters the right atrium of the heart to eventually be pumped to the lungs for oxygenation and return to the tissues. A *portal system* may be defined as beginning in a set of capillaries (i.e., in the gut wall) and ending in a set of capillaries (the sinusoids of the liver are classified as capillaries) *without* going through the heart. The hepatic portal vein carries about 75% of the total blood to the liver. The remaining 25% is carried by the *hepatic artery proper,* one of the terminal ramifications of the celiac trunk.

The hepatic portal vein normally also receives the *gastric veins (left gastric vein, coronary vein)* from the stomach. These veins have important anastomoses with *esophageal veins* in the walls of the lower portion of that organ (Fig. 5–1). The esophageal veins also have connections with the *azygous system of veins* (which drains into the superior vena cava) and with the *vertebral system of veins.* The latter is a valveless network of thin-walled veins in and around the vertebral canal; it possesses numerous connections with veins of the pelvis, abdomen, thorax and even the intracranial cavity and is important (due to the ease of reversal of blood flow) in the spread of tumor cells and/or infections. Portal hypertension causes a reversal in the normal direction of blood flow in the gastric veins, which results in the formation of varicosities (or varices) in the thin-walled esophageal veins. If the pressure is not too great in these vessels, this anastomosis forms an important collateral pathway (one of four) for the bypass of blood around the liver obstruction. Rupture of these esophageal varices, however, in our patient resulted in both the bleeding from the mouth and bloody stools.

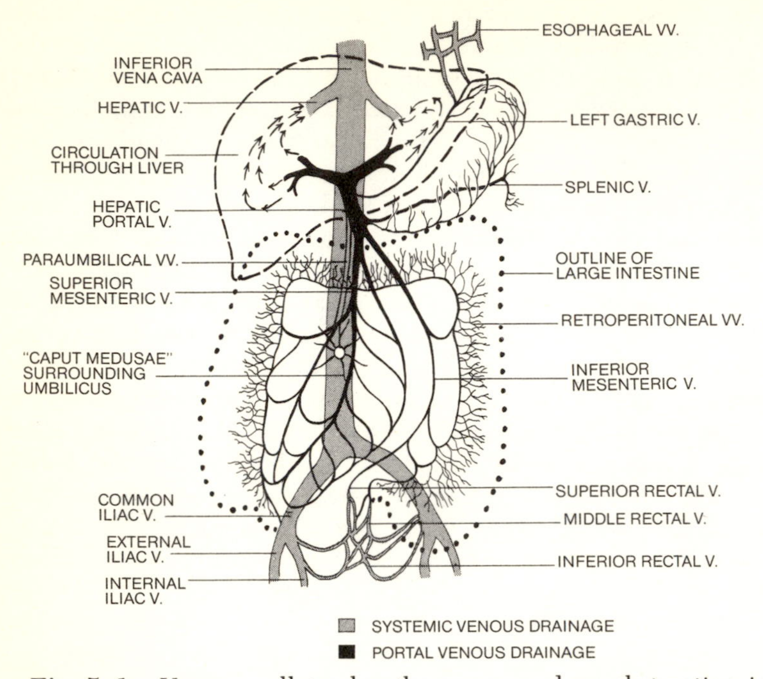

Fig. 5–1.—Venous collateral pathways around an obstruction in the liver (hepatic portal hypertension). This condition is characterized by a *"caput medusae"* (connections with *paraumbilical veins*), esophageal varices (connections with gastric veins) and hemorrhoids (connections with the *inferior mesenteric vein*). In addition, *retroperitoneal veins* allow blood to bypass the obstruction by flowing in a retrograde direction to systemic veins of the posterior abdominal wall. (Redrawn from Crafts, R. C.: *A Textbook of Human Anatomy*. Copyright © 1966, The Ronald Press Co., New York.)

A second anastomotic channel is that between the *superior rectal tributaries* of the *inferior mesenteric vein* and the *middle* and *inferior rectal veins* (see Fig. 5–1). Blood forced in a retrograde direction down the inferior mesenteric vein due to increased pressure in the portal vein will overload the rectal (hemorrhoidal) system of veins, resulting in clinically detectable hemorrhoids, as was observed in this case. Blood from the middle and inferior rectal veins drains to the internal and common iliac veins and thus to the inferior vena cava. This is, therefore, a second pathway by which blood may bypass the liver on the way to the heart. In addition, blood may reach the vertebral plexus via collateral connections with the tributaries of the internal iliac veins and ascend or descend depending on changes in intrathoracic and intra-abdominal pressures. The accumulation of ascites fluid frequently observed in cirrhosis is the result of increased pressure in both the inferior and superior mesenteric veins. As a result, fluid is forced out of the veins and venules into the peritoneal cavity (hepatic edema).

A third way that venous blood may bypass the liver in obstructive disease is to travel in a retrograde direction in the *paraumbilical veins*, which are small tributaries of the hepatic portal vein (see Fig. 5–1). The paraumbilical veins occupy a position near the free edge of the falciform ligament and form connections with small subcutaneous veins located circumferentially around the umbilicus. Retrograde blood flow in this case results in enlargement of the latter veins, which form a characteristic radiating pattern with the navel as the axis; this phenomenon has been termed the *caput medusae* due to its resemblance to a crown of snakes. These radiating veins communicate with bilateral venous channels known as the *thoracoepigastric veins*, which connect the *lateral thoracic veins* superiorly with the *superficial epigastric veins* inferiorly. Ordinarily, the thoracoepigastric veins are insignificant or invisible, although they are normally prominent in some animals, e.g., the horse. The venous anastomoses in this example, as in others, occur usually on a capillary basis but have great capacity for slow expansion under such conditions as cirrhosis of the liver. Venous blood from the superior part of the trunk surface drains to the axillary vein and thence to the superior vena cava; that from the inferior portion drains to the external iliac vein and from there to the inferior vena cava.

The fourth and final way that blood may circumvent the liver in such cases is through connections between *retroperitoneal veins* and tributaries of the inferior vena cava. This is the one case in which readily detectable clinical symptoms are absent, primarily due to the small size of the vessels and the fact that the anastomoses occur deep in the tissues of the posterior abdominal wall. These anastomoses occur anywhere that an organ or portion of an organ (i.e., liver, duodenum, ascending and descending colon) comes in direct contact with tissues of the body wall without any peritoneum interspersed. Due to the relatively large area of tissue involved, this example of collateral circulation probably allows a greater volume of blood flow around the liver than any of the other three. The blood flows to phrenic and lumbar veins and thus to the inferior vena cava.

The malnutrition observed in the patient was probably the result of faulty eating habits associated with alcoholism as well as of the fact that his liver was functioning improperly. His frequent lapses into a comatose state were related to a loss of blood and to his generally poor condition. Cirrhosis was believed for a considerable time to be an indirect result of alcoholism, i.e., the poor diet characteristic of the chronic alcoholic was thought to be responsible for the pathologic changes in the liver. This attitude has changed somewhat to the current belief that alcohol itself is the pathogenic agent, but there is still controversy on this point.

In the event that the problems associated with this condition become life threatening, e.g., the danger of hemorrhage from esophageal varices, a number of operations have been devised to shunt blood artificially around the liver. Basically, these procedures consist of surgically creating a direct anastomosis between the hepatic portal vein and either the inferior vena cava or the left renal vein. Patients are surprisingly able to tolerate such operations, presumably because the liver retains a satisfactory blood supply from the hepatic artery proper. In addition, the liver is capable of regeneration, diseased or destroyed cells being replaced through mitosis of healthy cells.

CASE 5–2: INGUINAL HERNIA

History

A 17-year-old boy who was in otherwise excellent health suffered an excruciating pain in the right groin area while trying to push a stalled automobile up a slight incline. He felt a painful bulge about midway between the midline and the anterior superior iliac spine, which his inspection showed to course medially and then turn inferiorly into the scrotum. The latter was greatly increased in size, especially on the right side. He began shortly to experience considerable abdominal pain, some perceived just deep to the point of origin of the bulge and other, more diffuse discomfort felt deeply in the abdominal cavity. In addition, he began to have abdominal cramps and became nauseated and lightheaded. He was able to flag down a passerby who took him to the emergency room of the nearest hospital. Following a brief examination, he was immediately scheduled for surgery. The problem was corrected and he was released from the hospital 6 days later. No recurrence or evidence of the problem was detectable in follow-up examinations.

Questions

Identify the specific clinical problem suffered by the young man.

Describe the relevant anatomy and the grounds for classification of clinical situations such as this.

What is your rationale for this diagnosis based on your knowledge of this area?

What body layers would surround the bulge as it proceeded toward the scrotum?

Why do you think some pain was felt locally (i.e., at the site of the bulge), whereas other pain was perceived deep in the abdominal cavity?

Would this be a congenital or acquired condition?

Based on your knowledge of the anatomy of this region, how do you think a surgeon would repair the defect?

Discussion

The condition described in this case is an *indirect inguinal hernia* brought about by a weakness in the anterior abdominal wall of the patient at the point where the spermatic cord begins its exit from the abdominal cavity. In general, a hernia involves the abnormal protrusion of part or all of a viscus or other structure through an orifice or into a recess of the body. In this case a loop of intestine was forced (due to the increased intra-abdominal pressure that resulted from straining to push the car) through the *deep inguinal ring*, down the *inguinal canal*, out of the *superficial inguinal ring* and into the right half of the *scrotum*, which contains the right testis.

The inguinal region or groin is that area between trunk and lower limb that is demarcated by the *inguinal ligament*. The latter is the inferior portion of the aponeurosis of the external oblique muscle, which extends between the pubic tubercle and the anterior superior iliac spine (Fig. 5–2). The ligament curves backward and then slightly superiorly to form a sort of shelf, the medial segment of which supports the spermatic cord during a portion of its course to the scrotum. Near the medial limit of the inguinal ligament is the superficial inguinal ring, an opening in the external oblique aponeurosis for the exit of the spermatic cord. The deep inguinal ring is another opening in the anterior abdominal wall, this one created by a regional deficiency in the internal oblique and transversus abdominis muscles. It is important to note that the deep ring is located *lateral* to the superficial ring, i.e., about midway between the pubic tubercle and the anterior superior iliac spine and slightly above the inguinal ligament. Thus, whereas there are two basic deficiencies in the anterior abdominal wall, they are out of register, therefore not forming a single anteroposterior defect. The superficial ring is strengthened posteriorly by the arching fibers of the internal oblique and transversus abdominis muscles (the *conjoined tendon* or *falx inguinalis*), whereas the deep ring is strengthened anteriorly by the external oblique aponeurosis. The oblique pathway between the two rings that transmits the spermatic cord from the abdominal cavity to the testis is the *inguinal canal*.

The inguinal region is divided for descriptive purposes by the *inferior epigastric artery*, a branch of the external iliac (see Fig. 5–2). This artery forms the lateral boundary of the *inguinal (Hesselbach's) triangle*. The medial boundary of this triangle is formed by the lateral border of the rectus abdominis muscle, whereas the inferior boundary is formed by the inguinal ligament. A hernia that enters the inguinal triangle is called a *direct inguinal hernia*. This type of hernia usually bulges forward to the superficial inguinal ring, but it rarely traverses the ring or enters the scrotum.

A second type of hernia enters the inguinal canal *lateral* to

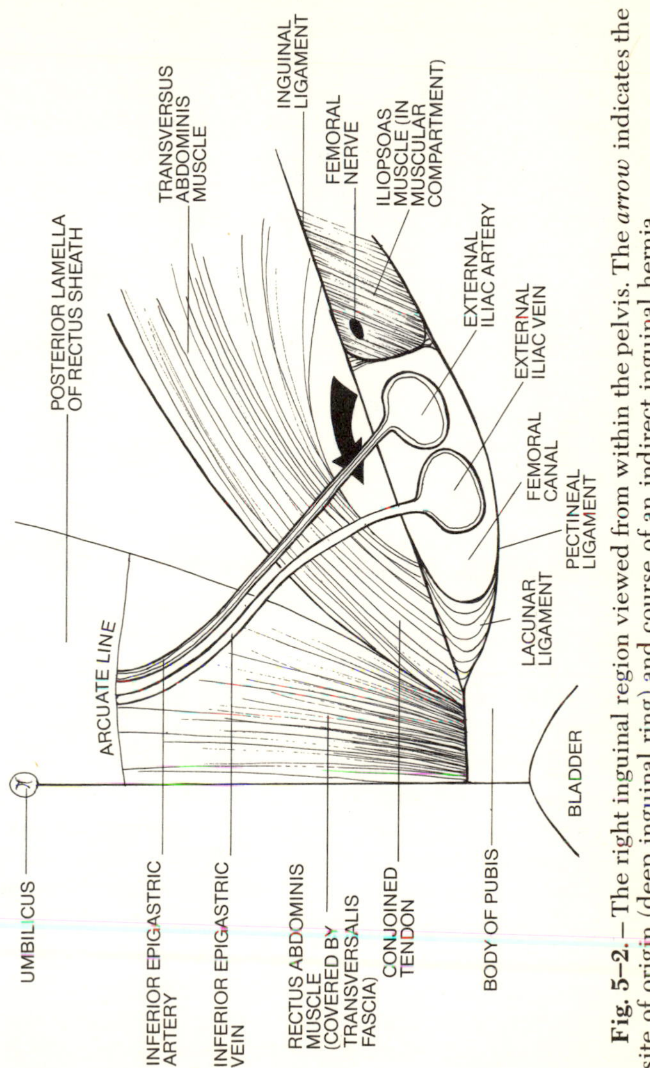

Fig. 5–2.—The right inguinal region viewed from within the pelvis. The *arrow* indicates the site of origin (deep inguinal ring) and course of an indirect inguinal hernia.

the inferior epigastric artery, i.e., through the deep inguinal ring, and is referred to as an *indirect inguinal hernia.* This type is more common than the direct, occurs more frequently on the right side and is more common in men than in women.

A third type of hernia that occurs in this area is the *femoral hernia.* In this case the loop of intestine or other tissue (sometimes fat) protrudes behind the inguinal ligament into that fat-filled space between the lucunar ligament and the femoral vein (i.e., the *femoral canal*). If severe, the herniated material may create a visible lump in the thigh in the medial area of the femoral triangle. Femoral hernia occurs more frequently in adults than in children, in women than in men and on the right side than on the left.

In this patient, as in most, the hernia was easily diagnosed as an indirect inguinal because the bulge originated near the middle of the inguinal ligament, i.e., lateral to the inferior epigastric artery, and because it traversed the inguinal canal, exited from the superficial inguinal ring and entered the scrotum. An indirect inguinal hernia often is associated with a vestigial diverticulum of the peritoneum known as the *processus vaginalis.* In fetal life this tubular outgrowth of peritoneum extends into the scrotum. The testis descends along its posterior surface into the scrotum and then invaginates and receives a covering, the *tunica vaginalis testis,* from the peritoneal tissue. The remainder of the peritoneal tube (i.e., the portion between the testis and the abdominal wall) ordinarily closes off and degenerates into a fine fibrous cord, leaving only a fine or usually imperceptible scar on the interior of the abdominal cavity. If this does not happen, however, the opening of the tube, which may persist until adulthood, forms an open invitation to the entering of a loop of intestine, as occurred in this case.

If the loop of intestine reaches the scrotum, the layers of tissue that cover it will be the same as those ordinarily found covering the spermatic cord. Besides the peritoneum, which is usually not demonstrable in the spermatic cord, the layers from within outward are as follows: (1) The *internal spermatic fascia.* This is the name given to the transversalis fascia (the deep fascia of the transversus abdominis muscle) in the spermatic cord, which is picked up as the herniated loop traverses the deep inguinal ring. (2) The *middle spermatic fascia* or *cremasteric fascia,* which is derived from the deep fascia of the internal oblique muscle. Interspersed in this fascia are muscle fibers, the *cremasteric muscle,* which are derived from the *internal oblique muscle* and which extend on the cord into the scrotum. (3) The *external spermatic fascia,* a prolongation of the deep fascia of the external oblique muscle. This layer is added to the loop of intestine as it traverses the superficial inguinal ring. When a preformed outpocketing of peritoneum exists (as in a persisting patent processus vaginalis), the indirect inguinal hernia is spoken of as "congenital." If a sac is formed de novo by the herniated material, it is known as "acquired."

In the patient the pain at the deep inguinal ring was the result of a strangulation of the gut loop, causing a restriction or loss of blood supply. Because this restriction can result in gangrene of the intestine in a very short period of time (12 hours or less), it is the cause for a medical emergency. The nausea (vomiting is frequently evident) experienced by the patient resulted from obstruction of the gut and interference of normal peristaltic activity caused by the strangulation at the deep inguinal ring. The deeper, more diffuse pain resulted from pressure caused by the herniated material on the testis itself. It should be remembered that the testis develops within the abdominal cavity and then descends, carrying its nerve supply, through the inguinal canal and into the scrotum. Pain from the testis is therefore largely referred to the abdominal cavity and is perceived as a sickening sort of sensation, resembling severe abdominal cramps.

To test for possible weakness in the inguinal canal, the physician during a routine examination of the male places the tip of the finger in the superficial inguinal ring and asks the individual to cough. The latter maneuver increases the intra-abdominal pressure and, if a tendency toward herniation exists, the physician is likely to feel a transient bulge at the superficial ring. Surgical repair of a hernia consists of returning the herniated material to the abdominal cavity, of ligating the opening to a peritoneal sac, if one exists, and of reducing

THE SUICIDE CLAUSE

Susan Vanost Silberman

All three of them look at their bowls with dismay. Amanda, of course, is the one to say it. "Macaroni and cheese? Again?"

"I love macaroni and cheese," Matt says, hardly mustering enough enthusiasm to sound convincing. He dips his spoon slowly into the orange mass and brings it to his mouth. That only makes it worse. I smile at him, blinking away my tears.

I feel the same way they do. As hungry as I am, I almost gag on the first bite. No wonder people complained about Manna from Heaven. It may be a blessing to have food at all, but the stomach craves variety. I get paid on Friday—maybe then we can go to McDonalds.

But we're almost out of toilet paper. How can anyone live without toilet paper? There's half a roll left for the four of us until Friday. I can't tell the kids not to use toilet paper. Last night I used my underwear before dropping it in the hamper. Maybe I can get a roll from work, but imagine if I get caught stealing toilet paper.

the size of the openings and the canal through which material coursed. In the case of an indirect inguinal hernia, the deep ring usually is made surgically smaller by suturing the arching borders of the transversus abdominis and internal oblique muscles to the inguinal ligament or to the pectineal ligament.

CASE 5–3: ESOPHAGEAL DISORDER

History

A 53-year-old woman had suffered for many years from a vague discomfort she defined as "heartburn." She described her symptoms as a burning sensation in the "pit of the stomach," often radiating superiorly to the area behind the sternum and sometimes extending as far superiorly as the throat. Occasionally, this would manifest itself as what she called "sour stomach," a regurgitation or reflux of the stomach's contents into the esophagus, usually leaving a sour taste in the mouth. These symptoms were considerably aggravated by postural changes, specifically when she bent over at the waist from a standing position or occasionally upon lying down. In general her condition was improved with the aid of antacids. Of late, however, she had begun to experience dysphagia (difficulty in swallowing), and she developed a chronic cough that frequently awakened her during the night when she was in the recumbent position. The cough was relieved somewhat if she slept with her head considerably elevated or in a sitting position. She was prompted to visit her physician, however, after observing recurrent dark granular stools followed by a crisis in which she vomited a large amount of blood. She was examined radiographically and with an esophagoscope. The former examination revealed a small peritoneal sac, which contained a small amount of the cardia of the stomach protruding through the esophageal hiatus in the diaphragm into the thoracic cavity. The latter demonstrated a stricture of the esophagus at the esophagogastric junction and a bloody, weeping esophageal mucosa. An esophageal balloon was inserted and inflated to arrest the bleeding, and surgery was performed soon thereafter to correct the defect in the diaphragm as well as to resect the esophagus, removing the small segment with the stricture. The patient's recovery was complete, and she was totally relieved of her symptoms postoperatively.

Questions

Was there more than one condition bothering this woman? If so, identify them, and discuss the symptoms and/or pathologic changes brought about by each.

Describe the anatomy of the esophagus, especially with regard to its junction with the cardiac region of the stomach.

What demarcates the transition from esophagus to stomach? Include in your discussion a description of the esophageal opening (hiatus) in the diaphragm.

What major structure(s) have to be protected during surgical operations in this area?

Discussion

The woman was diagnosed as suffering from *hiatus* (or *hiatal*) *hernia* compounded by gastroesophageal reflux (regurgitation of stomach contents into the esophagus). Hiatus hernia is defined as the abnormal protrusion of any structure (usually a portion of the stomach) through the esophageal hiatus in the diaphragm. A hiatus hernia may or may not cause gastroesophageal reflux, and gastroesophageal reflux may occur in the absence of hiatal hernia. It is probably safe to say that all adults at one time or another in their lives have suffered from gastroesophageal reflux. In this case, the fact that the patient experienced aggravation of her epigastric burning pain and regurgitation of gastric contents as a result of postural changes (stooping or lying) assures the diagnosis of symptomatic esophageal reflux.[1] Radiograms confirmed the presence of a hiatus hernia (Fig. 5–3). Therefore, although the two conditions may occur independently of each other, it appears likely that they were closely related in this patient. Thus, the gastroesophageal reflux was the result of the hernia (see below), and the esophageal stricture (the cause of the dysphagia) and the bleeding were the result of the chronic reflux. The latter resulted in inflammation of the esophagus, or *esophagitis*. Incidentally, the patient's nocturnal cough was the result of aspiration of gastric contents into the lungs, one of the most serious complications of gastroesophageal reflux.

The esophagus begins at the level of C. V. 6. It descends in the superior mediastinum in the median plane posterior to the trachea and then in the posterior mediastinum where it lies immediately behind the pericardium lining the left atrium. It begins to deviate to the left just inferior to the bifurcation of the trachea and then passes through the diaphragm at the level

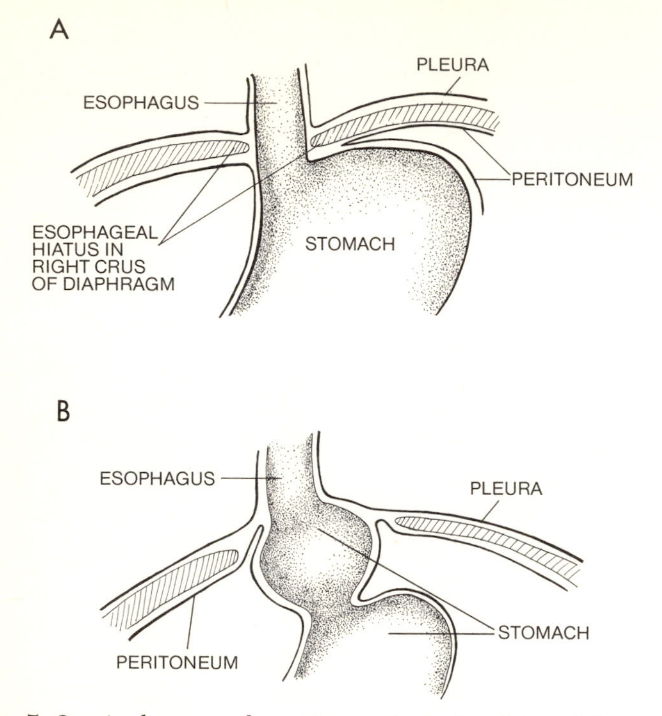

Fig. 5–3. – **A,** the normal position and relationships of the *esophagus* and *stomach* to the diaphragm. **B,** a type I (sliding) hiatal hernia (based on Edwards, *et al.*[2]).

of T. V. 10. Portions of the cervical and thoracic vertebral column lie posterior to it throughout its course. The junction of the esophagus with the cardia of the stomach lies approximately 2 cm to the left of the midline and 1–2 cm below the esophageal hiatus in the diaphragm. The intra-abdominal portion of the esophagus is clothed by peritoneum on its anterior and lateral aspects; posteriorly, it lies in direct contact with the diaphragm.

The terms *cardia* and *gastroesophageal junction* often are used interchangeably; they connote the esophageal orifice of the stomach. According to Skinner,[1] three definitions of this junction have been proposed: "(1) The junction of squamous and columnar epithelium, (2) the point at which the tubular esophagus enters the pouch of the stomach, and (3) the junction between the esophageal inner muscle layer and the innermost layer of gastric muscle, the oblique or sling fibers." He believes that the most acceptable definition for clinical purposes is the epithelial or mucosal transition, which usually can be observed grossly with an esophagoscope. Microscopically, the esophageal submucosal mucous glands are not found below this point, and the cardiac glands of the upper portion of the stomach are not found above it. It is the general consensus of investigators who have studied the gastroesophageal junction that no anatomical intrinsic muscular sphincter can be demonstrated. On the other hand, there is general acceptance from physiologic, radiographic and esophagoscopic studies that a physiologic sphincter, which ordinarily prevents gastroesophageal reflux, exists at the junction.

Most workers agree that the esophageal hiatus is formed largely from the *right crus* of the diaphragm, although the left crus may, and frequently does, contribute some fibers to it (see Fig. 5–3). The hiatus is described as possessing right and left limbs, usually with clearly demonstrable margins. The right and left limbs usually are supplied by the right and left phrenic nerves, respectively. The opening transmits the vagus nerves as well as the esophagus; all of these structures naturally must be protected during surgical intervention in this area. The fibers of the esophageal hiatus form a muscular loop or sling, which creates a sharp angle between the esophagus and the greater curvature of the stomach. According to Edwards *et al.*,[2] this angle serves the following two functions: (1) it creates "an external and most significant sphincter which normally prevents reflux of gastric contents," and (2) it forces the cardia "to adopt a position away from the long axis of the esophagus, so that intra-abdominal pressure pushes the stomach against the left dome of the diaphragm rather than through the hiatus." These authors noted that this cardioesophageal angulation is lost in hiatus hernia.

According to Skinner,[1] hiatus hernia may be classified as follows: (1) Type I, in which the gastroesophageal junction is herniated into the thorax and is the proximal or leading point of the hernia. This also is known as an axial or sliding hernia and is the most common type (indeed, it is "now recognised as one of the most common abnormalities of the gastrointestinal tract"). Gastroesophageal reflux may be present; this is the type of hernia that was present in our patient. (2) Type II, in which the gastroesophageal junction remains fixed posteriorly at the hiatus, but the fundus or anterior wall of the stomach herniates through the hiatus. This type of hernia (also known as a rolling or parahiatal hernia) is much less frequently observed than type I, and is not usually accompanied by gastroesophageal reflux. (3) Type III, in which both type I and type II hernias are present. Both reflux and mechanical complications may accompany type III hernias. (4) Type IV, in which an organ or organs other than the stomach (e.g., transverse colon, small intestine, spleen) may herniate through the esophageal hiatus. Complications in this type of hernia depend largely on the organ involved.

Surgical repair of hiatal hernias consists basically of reducing the hernia and repairing the defect in the hiatus to prevent recurrence. Operations of this sort frequently take advantage of the relatively thickened endoabdominal fascia that surrounds the gastroesophageal junction; this tissue has been referred to as the *phrenoesophageal membrane* or *ligament* by some authors.

CASE 5–4: CHRONIC GASTRIC ULCER

History

A 43-year-old male executive underwent surgery for a chronic gastric ulcer. The procedure, which involved cutting the anterior and posterior vagal trunks in order to reduce gastric secretion, is termed a *vagotomy*. During the operation the surgeon accidentally incised an artery that occupied a position along the upper portion of the lesser curvature of the stomach. Examination revealed that this artery was considerably larger than that which ordinarily occupies this position. The surgeon immediately ligated the artery proximal to the injury and continued with the operation. Within a few minutes, however, he observed that the entire left half of the liver had become ischemic (deficient in blood supply). The ligature on the bleeding artery was removed in an attempt to prevent liver necrosis but, as a result, the patient hemorrhaged to death internally.

Questions

Which artery was incised during the operation?

Review the normal pattern of blood supply to the stomach and liver.

Why did ligation result in ischemia to the left half of the liver?

How do you explain the observation that the injured artery was considerably larger than normal?

What anastomoses exist with the artery in question that would provide collateral circulation to its area of supply in the event of a slowly developing occlusion?

Discussion

The vessel that was cut and subsequently ligated was the *left gastric artery,* one of the three major subdivisions of the *celiac trunk.* The latter is the first unpaired visceral branch of the abdominal aorta, arising just as the aorta passes between the crura of the diaphragm. The celiac trunk passes anteriorly through the dense matted substance of the celiac ganglia and plexus and, after a course of between only 1–3 cm, divides into the *left gastric artery,* the *common hepatic artery* and the *splenic artery* (Fig. 5–4). The following description of these vessels is the most commonly encountered in practice, but the student should be aware that great variability exists in this system.

The *left gastric artery* runs upward and to the left behind the peritoneum, forming a peritoneal ridge that is known as the *left gastropancreatic fold.* At the lesser curvature of the stomach, just inferior to the cardia, it obtains a position between the anterior and posterior layers of the lesser omentum and then turns downward to run between these two layers, following the concavity of this border of the stomach. From this position, it gives numerous branches to the anterior and posterior surfaces of the stomach.

The *common hepatic artery* passes to the right along the upper border of the pancreas, ridging the peritoneum to form the *right gastropancreatic fold.* Just above the first part of the

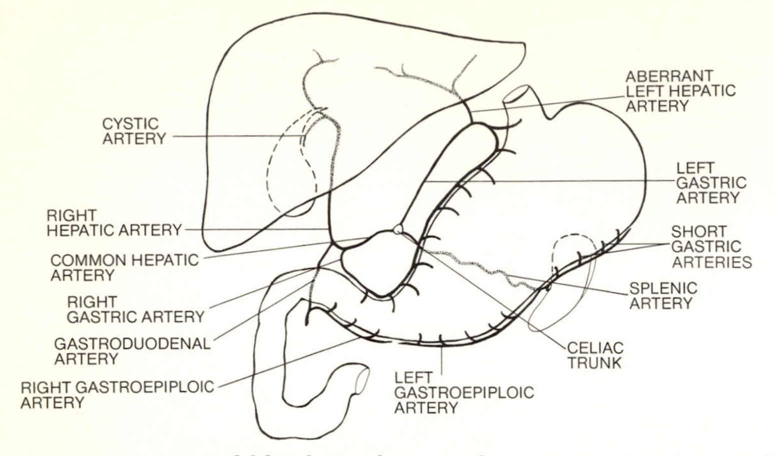

Fig. 5–4.—Arterial blood supply to the liver, stomach, spleen and first part of the duodenum. Note the *aberrant left hepatic artery* arising from the *left gastric artery.*

duodenum, it divides into the *hepatic artery proper,* the *right gastric artery* and the *gastroduodenal artery.* The hepatic artery proper ascends at the free edge of the lesser omentum (between the two layers of peritoneum) where it lies anterior to the hepatic portal vein and to the left of the bile duct. Near the liver it ordinarily divides into *right* and *left hepatic arteries* (sometimes there is a middle hepatic artery also), which supply the right and left halves of the liver, respectively. The right gastric artery enters the lesser omentum and courses to the left along the lesser curvature of the stomach. The gastroduodenal artery descends behind the first part of the duodenum, giving branches to that structure and to the head of the pancreas. It also gives off the *right gastroepiploic artery,* which runs to the left between the layers of the greater omentum at its attachment to the greater curvature of the stomach. It supplies branches to the anterior and posterior surfaces of the stomach and to the greater omentum.

The *splenic artery* is the largest branch of the celiac trunk. It passes to the left, usually in a very tortuous fashion, along the upper border of the pancreas to which it gives numerous branches. Near the spleen, it divides into several *splenic branches*, the *short gastric arteries* and the *left gastroepiploic artery*. The short gastrics that often arise from one of the terminal splenic branches run upward and medially to the greater curvature of the stomach where they divide into anterior and posterior branches to supply the respective surfaces of the stomach. The left gastroepiploic, which also may arise from one of the splenic branches, proceeds to the right between the layers of the greater omentum at the greater curvature of the stomach. Like its counterpart, it also gives branches to the anterior and posterior surfaces of the stomach and to the omentum.

Variations are so common in these blood vessels that even the most common pattern occurs in a minority of cases, i.e., less than 50% of the time.[3] One of the most frequent variations is that which occurred in this case, that of the *left hepatic artery* arising from the *left gastric artery*. This explains the extraordinarily large size of the left gastric artery as well as the ischemia produced in the left lobe of the liver upon its ligation. Occasionally, the entire hepatic artery proper arises from the left gastric. These vessels also may arise from the superior mesenteric artery. Morris[4] discusses these *aberrant* hepatic arteries as being either *replacing*, i.e., a substitution for the normally occurring branch, or *accessory* (in addition to one that is normally present). In this case the aberrant left hepatic artery was obviously replacing. This source notes, "Some sort of aberrant hepatic artery, either replacing or accessory, occurs in approximately 42% of individuals." This figure is large enough to cause obvious concern to surgeons operating in the area.

A rapid occlusion of this artery such as would occur in ligation or clamping would almost invariably produce ischemia and eventual death of the tissues supplied by it. A slowly developing occlusion could be tolerated, however, due to anastomoses with other arteries capable of providing collateral pathways when allowed to enlarge gradually. The two major sources of circumventing an occlusion of the left gastric artery are as follows: (1) *esophageal arteries*, branches of the thoracic aorta, which anastomose with branches of the left gastric around the cardia of the stomach; and (2) *short gastric arteries,* branches of the splenic artery, which cross the anterior and posterior surfaces of the stomach to anastomose with like branches of the left gastric. In addition, branches of the gastro-epiploic arteries may enter into anastomotic association with branches of the left gastric, and terminal branches of the right gastric artery may join those of the left gastric in a variable fashion along the lesser curvature of the stomach.

CASE 5–5: OBESITY

History

A 47-year-old businessman suffered from extreme obesity. During his most recent physical examination, he had to be weighed on the hospital's freight scale (his weight was 497 pounds). He experienced high blood pressure, dyspnea and bouts of dizziness. It was determined that his obesity was simply the result of overeating rather than some defect in metabolism. During the past decade he had attempted nearly every method of potential weight loss available including various diets and even hypnosis, but such attempts were rarely successful for any period of time. He would lose a few pounds, but then some problem would occur, usually associated with business pressures, and his food intake would increase once more. His physician had warned him that unless he could bring his weight down to under 200 pounds, his life would be in jeopardy. Since it was now apparent that the man could not or simply would not adhere to any program requiring self-control, he was scheduled for an operation that involved the "functional" removal of a portion of the small intestine. The procedure was successful, resulting in a slow, steady decline in weight, reaching a plateau at approximately 190 pounds.

Questions

How is this operation performed and what is its rationale?

What are the anatomical differences between the large and small intestines and among the various portions of the small intestine that might assist in their identification during surgery?

Describe the blood supply to the small intestine (duodenum excluded).

Discussion

In very general terms, this operation involves sectioning the small intestine, usually the ileum (hence the term *ileal bypass*) and rejoining the remaining segments. The bypassed section may be left in situ, with its blood supply intact in the event that rejoining part or all of it might be necessary. The operation is performed in order to reduce the absorptive surface of the small intestine and thereby bring about weight loss without reducing food intake commensurately. It usually is employed only when the more conventional methods of weight reduction have failed and the patient's life may be in danger, as in this case. The degree of weight loss apparently is related to the amount of intestine bypassed; too little may result in insufficient loss of weight, whereas the converse may occur if too much is removed.

The *large intestine* may be distinguished from the small intestine by the presence of the following characteristics: (1) *haustra* or *sacculations* (repeating dilatations separated from each other by constrictions); (2) *teniae coli*, three bands of longitudinal smooth muscle located approximately equidistant from one another, rather than a solid coat as occurs in the small intestine; and (3) *appendices epiploicae*, tabs of fat enclosed within peritoneum and suspended from the outer surface of the large intestine. The *small intestine* consists of the duodenum, jejunum and ileum from proximal to distal. The *duodenum* is about a foot in length and is largely retroperitoneal. Of the remaining portion of the small intestine (that suspended by the *mesentery*, i.e., the highly coiled portion), the *jejunum* comprises the proximal 40% and the *ileum* the distal 60%. This portion angles from the upper left abdominal quadrant at the *duodenojejunal flexure* to the lower right quadrant in the area of the right iliac fossa where the ileum joins the large intestine at the *ileocecal junction*. This is the line of attachment of the mesentery to the posterior abdominal wall. It is important to recognize that the coils of the small intestine are highly mobile and may be found almost anywhere within the abdominal cavity and commonly extend inferiorly into the pelvis. The *transverse colon* and *sigmoid colon* (both parts of

the large intestine) also are suspended by mesenteries and therefore are highly mobile and may occupy variable positions within the abdominal cavity.

The morphologic differences between the jejunum and ileum are subtle, especially the closer the approach to the dividing point between the two. Actually, the only accurate means of identifying jejunum and ileum in this transitional area is with the aid of histologic technics and microscopy. There are discernible differences, however, between the proximal end of the jejunum and the distal end of the ileum. The jejunum usually has less material within its lumen, thicker walls and a higher degree of vascularity, and the adjacent mesentery contains less fat than is apparent in the portion that suspends the ileum.

The *superior mesenteric artery* courses between the layers of the mesentery, giving numerous *jejunal* and *ileal branches* to the respective portions of the small intestine. These branches join to form anastomotic loops or *arcades*. At the more distal and mobile portions of the small intestine, the number of tiers of arcades increases so that in the ileum there may be as many as four of five tiers (Fig. 5–5). It is thought that the arcades are related to mobility in the sense that the gut may be stretched or twisted with little or no effect on the blood vessels. In other words, as a section of intestine is flattened out, for example, a single longitudinally oriented artery might be ruptured, but the tiers of arcades are each flattened only minimally. From the terminal arcade, *straight arteries* or *vasa recta*

Fig. 5–5. – Ileal arcades.

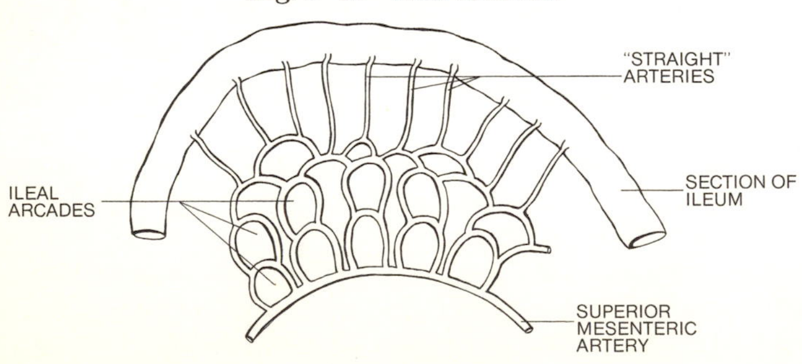

extend to the gut wall itself. These are functional *end arteries*, that is, there are insufficient anastomoses within the walls of the intestine to permit ligation without cutting off the arterial blood supply to that section. The veins of the small intestine accompany the arteries and form the *superior mesenteric vein*, which joins the *splenic* and *inferior mesenteric veins* in a variable fashion to form the *hepatic portal vein*. In this way, nutrients from the small intestine are carried to the liver for storage and metabolism.

CASE 5–6: LOWER GASTROINTESTINAL DISORDER IN AN INFANT

History

A 6-month-old boy weighed only seven pounds at the time of admission to the hospital. He was suffering from chronic constipation. distention of the abdomen and intermittent bouts of vomiting and diarrhea. The constipation could be controlled to some degree by enemas, but he rarely experienced what even approximated a normal bowel movement. Roentgenographic examination revealed a severely distended sigmoid colon. In the process of a digital rectal examination, great quantities of feces and gas were released. A biopsy of the wall of the affected portion of the large intestine revealed the absence of autonomic ganglion cells and plexuses. Surgical excision of the affected portion of the colon was performed and the cut ends of the bowel rejoined. This resection was successful, and the boy recovered normal bowel function.

Questions

What is the name of this condition and what is its cause?
Describe briefly the anatomy of the large intestine.

Discussion

This young child was suffering from *megacolon* or *Hirschsprung's disease*, which manifests itself as a distended section of large intestine that does not demonstrate normal muscular peristaltic action. This results in a retention of feces with the attendant symptoms and may therefore be considered a functional form of bowel obstruction. Turell[5] states that in this disease, Meissner's and Auerbach's plexuses lack ganglia in the affected segments, their place being taken by "whorls of nerve fibers." *Auerbach's plexus* also is known as the *myenteric plexus*, since it is located in the area between the two layers of smooth muscle in the gut wall. *Meissner's plexus* also is referred to as the *submucous plexus* because of its location in that layer of the wall. Ordinarily, small ganglia are found in these plexuses, which receive nerve fibers from the autonomic nervous system. The parasympathetic neurons end on the ganglia of the myenteric (muscular) and submucous plexuses, causing peristalsis as well as secretion. Sympathetic fibers appear to terminate directly on the muscle of the gut wall and are said to be inhibitory to the parasympathetic impulses. Absence of the ganglia, as occurs in this disease, results in a loss of tonus and peristalsis of the gut wall, with consequent retention of feces and distention. It should be recalled that stomach mobility and secretion may be inhibited by cutting the vagal supply to the stomach (vagotomy).

The large intestine consists of the *cecum* (a blind pouch located near the right iliac fossa that gives rise to the *vermiform appendix*), *ascending colon*, *transverse colon*, *descending colon* and *sigmoid colon*. The latter leads into the *rectum* and *anal canal*. Only the transverse and sigmoid colons possess a mesentery; the remainder of the large intestine is retroperitoneal. The blood supply is via the *superior* and *inferior mesenteric arteries*. The superior mesenteric gives off an *ileocolic artery*, which, as its name implies, gives branches to the terminal ileum, cecum and appendix; a *right colic artery* to the ascending colon; and a *middle colic artery* to the transverse colon. The inferior mesenteric gives off a *left colic artery* to the descending colon and *sigmoid branches* to that portion and then descends into the pelvis as the *superior rectal artery*. The various colic arteries join around the margin of the large intestine as the *marginal artery*. This is an effective anastomotic channel for collateral circulation in the event that one of the colic branches becomes blocked.

REFERENCES

1. Skinner, D. B.: Anatomy; Symptoms, in Skinner, D. B., Belsey, R. H. R., Hendrix, T. R., and Zuidema, G. D. (eds.): *Gastroesopha-*

geal Reflux and Hiatal Hernia (Boston: Little, Brown and Co., 1972).
2. Edwards, E. A., *et al.: Operative Anatomy of Thorax* (Philadelphia: Lea & Febiger, 1972).
3. Gardner, E., *et al.*: *Anatomy* (3d ed.; Philadelphia: W. B. Saunders Co., 1969).
4. Anson, B. J. (ed.): *Morris' Human Anatomy* (12th ed.; New York: McGraw-Hill Co., 1966).
5. Turell, R.: *Diseases of the Colon and Anorectum* (Philadelphia/London/Toronto: W. B. Saunders Co., 1969).

Chapter 6

URINARY SYSTEM

CASE 6–1: KNIFE WOUND IN THE KIDNEY

History

A young man was seriously injured in a gang fight when he received a deep knife wound in the right posterolateral side of the back. Specifically, the wound was at a point 2 cm inferior and about 4 cm medial to the tip of the twelfth rib, i.e., at a horizontal level that would pass through the body of L. V. 2. The man was rushed in a police ambulance to the emergency room of the county hospital where, due to severe hemorrhaging, the right kidney was surgically removed. Examination revealed that the knife blade had penetrated the kidney in a plane from posterolateral to anteromedial, emerging near the hilus. Peritonitis developed 3 days following surgery; this was detected and eradicated immediately with antibiotics. He was released from the hospital 10 days after injury.

Questions

Describe the posterior (lumbar, retroperitoneal) approach to the kidney. Is this the most common surgical technic for achieving access to the kidney?

Identify from without inward *every* layer or structure that was encountered (and therefore incised) by the knife blade as it penetrated the man's back.

Discuss the renal fascia and associated fat.

Describe the basic gross anatomy of the kidney and the structures that enter or leave it.

What structures would need to be cut and ligated in order to remove a kidney?

Why did peritonitis develop as a result of this injury?

Discussion

The posterior (or lumbar or retroperitoneal) surgical approach to the kidney is the one commonly used in kidney transplants and other manipulations of this organ. The reasons for this are that the kidney is readily accessible from behind and that the technic avoids complications that may arise from possible contamination of the peritoneal cavity in anterior approaches.

Anatomically, this approach to the kidney involves incising and reflecting the skin and subcutaneous tissue *(tela subcutanea)* from approximately T. V. 12 to L. V. 4 or 5. A vertical incision is made along the spines of these vertebrae, and two horizontal incisions are then carried laterally from the superior and inferior limits of the vertical one. This procedure exposes the underlying muscles, which are to be reflected in the following manner. The *latissimus dorsi* is incised at the origin of the muscle fibers from the broad, flat *thoracolumbar fascia.* The muscle fibers are reflected laterally and the fascia medially. This exposes the partially underlying *external* (abdominal) *oblique muscle,* the free edge of which may be traced upward to the tip of the twelfth rib. This rib is the key landmark to the exposure of the kidney from behind. The *serratus posterior inferior muscle,* although usually thin and poorly developed, occupies a position deep to the latissimus dorsi and may have to be removed. The external oblique may be cut and reflected laterally, which aids in the exposure of the underlying *internal* (abdominal) *oblique muscle.* This muscle is in turn incised and the cut ends reflected medially and laterally to expose the *transversus abdominis muscle* or its fascial origin. The above three abdominal wall muscles may be identified by the course and direction of their fibers. The external oblique fibers course downward and medially, the internal oblique upward and medially (at basically right angles to the former) and the transversus abdominis fibers run horizontally around the abdominal wall. The *subcostal nerve* (anterior primary ramus of thoracic T_{12}) may be observed emerging through the transversus abdominis muscle or its fascial origin and coursing laterally below the twelfth rib between this muscle and the internal oblique. The *iliohypogastric nerve* (derived from the first lumbar spinal nerve) is found running parallel and about 3–4 cm inferior to the subcostal nerve. An incision *between* these two nerves parallel to them and to the twelfth rib, through the remaining layers of the abdominal wall, will expose the posterior surface of the *kidney*.

Thus far then, the following structures were encountered by the knife on its way to the kidney (see Fig. 6–1): skin; subcutaneous tissue; and latissimus dorsi, serratus posterior inferior, internal oblique (note that the external oblique muscle was missed because the blade entered medial to a vertical line drawn inferiorly from the tip of the twelfth rib, i.e., the posterior free edge of that muscle) and transversus abdominis muscles (or the latter's fascial origin, depending on the relative degree of development). The kidney itself is directly surrounded by fat, known as the *perirenal (perinephric) fat*, sometimes referred to as the fatty capsule of the kidney. This latter term is not to be confused with the true fibrous capsule, which lines the exterior of the kidney itself. External to the

Fig. 6–1. – Relationship of the kidney to the layers of the posterolateral abdominal wall, to the fascia and fat that surround it and to the peritoneum. The *broken arrow* indicates the course of the knife wound described in Case 6–1. (Based on Grant, J. C. B.: *An Atlas of Anatomy* [6th ed.; Baltimore: Williams & Wilkins Co., 1972].)

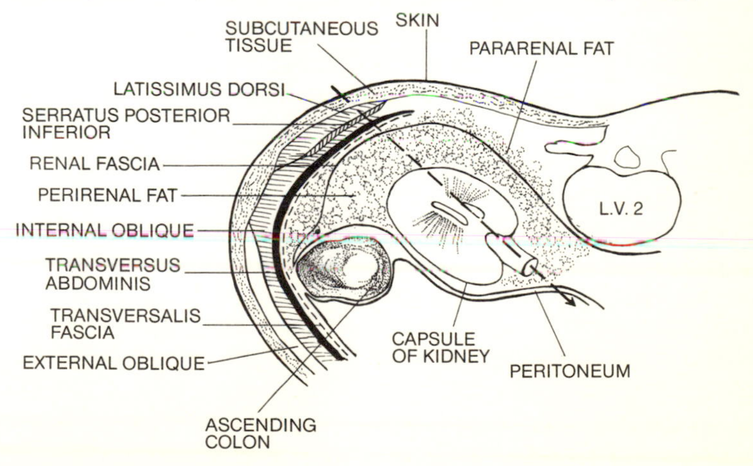

perirenal fat is a condensation of extraperitoneal connective tissue known as the *renal fascia,* which splits to enclose the kidney. The anterior and posterior lamellae of the renal fascia fuse superiorly over the superior pole of the kidney and enclose the suprarenal gland. Inferiorly, this fusion is weak or deficient and, as a result, repeated jolting actions to the body, as occur to jackhammer operators or some motorcycle riders, may cause inferior displacement of one or both kidneys. It should be noted that the primary support of the kidneys does not come from the renal fascia but, rather, from the blood vessels that enter and leave the organ. Occasionally a kidney may not be retroperitoneal but, rather, will be suspended by a mesentery (the so-called "floating" kidney). This is a potentially hazardous situation in that the kidney may twist on its blood supply, thereby cutting it off. External to the renal fascia is another condensation of fat, this layer called the *pararenal (paranephric) fat.* The remaining layer is the *transversalis fascia,* one of the components of the continuous sheet of *endoabdominal fascia.* This layer in other areas of the abdominal wall is called *psoas* (major) *fascia* or *quadratus* (lumborum) *fascia.* Thus, completing the list of structures from without inward, there are the transversalis fascia, pararenal fat, renal fascia, perirenal fat and the capsule of the kidney. It should be pointed out that the kidney lies partially against the *quadratus lumborum muscle* posteriorly and that this muscle usually will need to be retracted medially in order to visualize the full extent of the kidney.

The kidneys lie at the sides of the vertebral column along the lateral border and in the plane of the psoas major muscle. In the recumbent position, the kidneys lie between T. V. 12 and L. V. 3, whereas in the erect position they descend to lie between L. V. 1 and L. V. 4. The right kidney normally is slightly lower than the left. The kidneys may move vertically up to 7 cm during different respiratory phases.

A cut section of kidney demonstrates an inner *medullary* area and an outer *cortical* area. Grossly the medulla is composed of *renal pyramids,* which empty via *papillae* into the *minor calyces.* Two or more minor calyces join to form the total of two or three *major calyces,* which unite to form the *renal pelvis,* the upper expanded end of the *ureter,* which lies within the *renal sinus.* The cortex extends peripherally to the capsule and contains the *renal corpuscles* (Bowman's capsules plus the capillary glomeruli). Cortical tissue that extends into the medulla between the pyramids constitutes the *renal columns,* and medullary tissue that radiates from the base of the pyramids into the cortex is known as *medullary rays* (for more detail, see Case 6–3). The student should consult a histology textbook for a detailed description of the microscopic structure of the kidney.

The blood supply to the kidneys is derived from the *renal arteries,* branches of the abdominal aorta. Each artery usually divides into two branches, an *anterior* and a *posterior,* for supply to those portions of the organ. The anterior branch in turn usually divides into four branches, an *apical* and an *upper, middle* and *lower branch.* It should be pointed out that the kidney arterial supply is variable and not infrequently multiple. Polar branches may be given to both the superior and inferior poles, the upper one often from a suprarenal artery. The arterial branches to the kidney are segmental; that is, due to a lack of internal or external anastomoses, they are functional end arteries, and occlusion of one of the major branches results in the death of the segment ordinarily supplied by that branch. As the arteries enter the kidney at the hilus, they divide into *interlobar arteries,* which progress peripherally through the medulla. These give off *arcuate arteries,* which run at right angles to the former along the bases of the renal pyramids. The latter arteries in turn give off *interlobular arteries,* which run peripherally in the cortex. These eventually provide the *afferent arterioles* to the glomeruli. There are no anastomoses between the various arcuate arteries, but some capillary anastomoses are present in the cortex between ramifications of the interlobular arteries. These are not sufficient, however, to establish collateral circulation in the event of an acute occlusion of a major renal artery. The *renal veins,* which are less frequently variable, drain into the inferior vena cava. The left renal vein receives as tributaries, phrenic, suprarenal and gonadal veins and, therefore, drains a large area on the left side of the abdomen.

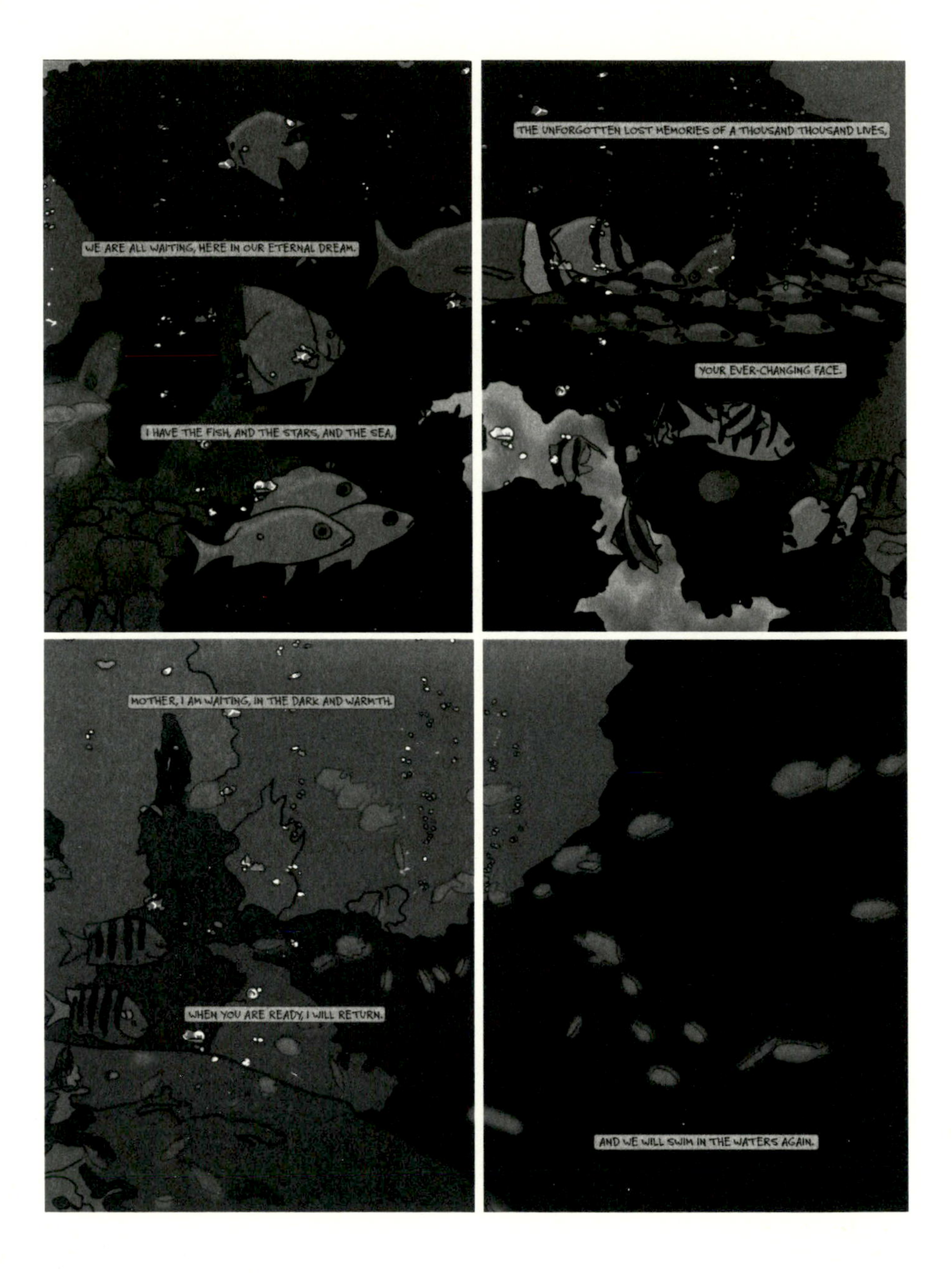
WE ARE ALL WAITING, HERE IN OUR ETERNAL DREAM.
I HAVE THE FISH, AND THE STARS, AND THE SEA,
THE UNFORGOTTEN LOST MEMORIES OF A THOUSAND THOUSAND LIVES,
YOUR EVER-CHANGING FACE.
MOTHER, I AM WAITING, IN THE DARK AND WARMTH.
WHEN YOU ARE READY, I WILL RETURN.
AND WE WILL SWIM IN THE WATERS AGAIN.

The nerve supply to the kidneys is still a subject of controversy. Postganglionic *sympathetic* neurons, arising from synapses in the aorticorenal ganglia, are probably entirely vasomotor in function. The presence or absence of vagal fibers is in doubt. *Parasympathetic fibers,* if they do exist, may derive from pelvic splanchnic nerves (S_{2-4}) due to the low origin and later ascent of the kidney. Lymphatic vessels from the kidney drain to lumbar nodes and trunks and, therefore, to the cisterna chyli and eventually to the thoracic duct.

In order to remove a kidney, the only structures that would need to be cut and ligated are those that constitute the so-called *renal pedicle,* i.e., the *renal artery, renal vein* and *ureter.* Just external to the hilus of the kidney, these structures usually occupy a position from anterior to posterior of vein, artery and ureter in that order, although this is not always the case. Attention to the possibility of arterial variation is extremely important in kidney surgery. An individual can live normally with only one kidney (the remaining one usually will hypertrophy), but loss of both of these organs is incompatible with life.

In our patient, peritonitis developed because the tip of the knife blade extended, although barely, into the peritoneal cavity (see Fig. 6 – 1 for the relationships of the kidney to this cavity). The knife did not enter the intestine, however, and therefore the bacteria responsible for the infection were carried into the cavity on the knife itself.

CASE 6–2: EXTRAVASATION OF URINE

History

A 12-year-old boy, while walking on a suspended two-by-four, lost his balance and fell with one leg on either side of the board. He came down hard on the edge of the two-by-four, most of the impact being absorbed in the area between the scrotum and the anus. He was in considerable pain and unable to walk for a brief period of time, but the pain gradually subsided to a dull ache and he proceeded slowly to his home. He described the incident to his parents who examined him and found that the skin was not broken, although a rather severe hematoma was forming subcutaneously. The injury did not seem serious enough to warrant contacting a physician. The following morning, however, the boy noticed his penis and scrotum were enlarged, turgid and reddish purple. In addition, he experienced painful urination and observed a considerable amount of blood in the urine. The parents immediately called their physician, who upon hearing their description of the symptoms had the boy admitted to a hospital. On examination a bloody subcutaneous exudate was observed that extended anteriorly from a straight line between the ischial tuberosities, along the ischiopubic rami and into the scrotum. Anteriorly the exudate continued into the lower part of the abdominal wall and extended superiorly about halfway between the pubic symphysis and umbilicus in fingerlike projections. Inferiorly it stopped on each side at a line parallel but slightly inferior to the inguinal ligament. The boy was treated surgically and, as a result, no additional exudate was formed or deposited. That which was there at the time of admission began to be reabsorbed soon after the operation. The boy was released 7 days after surgery and recovery was rapid and uneventful.

Questions

What structures were injured in the boy's fall?

Pinpoint as specifically as you can the location of the injury that resulted in the observed symptoms (i.e., the exudate and painful urination).

Describe the fascial layers in this area that were responsible for the limitation or facilitation of the spread of the exudate.

How would a surgeon repair this injury?

Discussion

The boy's fall on the two-by-four ruptured the *urethra* at the point immediately after it passes through the urogenital diaphragm into the penis (i.e., at the transition from the membranous portion of the urethra to the penile portion). At this point, the urethra makes a sharp bend anteriorly, and it is here that the structure is especially vulnerable to injury. Urine and the blood that entered it from the injured urethra (and perhaps from surrounding tissues that also may have been injured) were then free to infiltrate the area surrounding the rupture.

In order to understand how the extravasated urine and blood were confined to certain areas of the perineum while spreading somewhat freely to others, it is necessary to appreciate the attachments and relationships of the various fasciae in this region. Unfortunately, this is one of the most controversial and difficult areas to comprehend in the human body. This description, while maintaining accuracy, will be as simplified as is possible. The muscular *urogenital diaphragm,* which extends from near the pubic symphysis anteriorly along the ischiopubic rami posteriorly to the level of the ischial tuberosities, is covered above and below by deep fascia (Fig. 6–2, A). The *inferior fascia of the urogenital diaphragm* also is referred to as the *perineal membrane.* Along with the *superior fascia,* it forms the boundaries of the *deep perineal pouch* (space, compartment), which in the male contains the urogenital diaphragm and the bulbourethral glands. The urogenital diaphragm is pierced by the *membranous portion of the urethra.* Inferior to the perineal membrane are the bilateral *corpora cavernosa,* each covered by an *ischiocavernosus muscle* and the centrally located *bulb of the penis,* enclosed by the *bulbocavernosus muscle.* These muscles are covered inferiorly by the *deep perineal fascia* (also known as the external perineal fascia or inferior perineal fascia of Gallaudet). This layer is, in effect, the deep fascia of the ischiocavernosus and bulbocavernosus muscles. Between these muscles this fascia lies on the perineal membrane where the latter is exposed (see Fig. 6–2, B). The deep perineal fascia is continued onto the penis where the two crura (corpora cavernosa) and the bulb meet; here it is referred to as the *deep fascia of the penis (Buck's fascia).* This fascia is continuous on the anterior abdominal wall with the deep fascia on the outer surface of the external oblique muscle. Inferior (superficial) to this layer is the *superficial perineal fascia,* which, like superficial fascia in general, may be subdivided into an outer fatty layer and an inner membranous layer. This combined fascial layer is continuous with that in the anterior abdominal wall, which also contains an outer fatty layer *(Camper's fascia)* and an inner membranous layer *(Scarpa's fascia).* In the perineum the membranous layer often is referred to as *Colle's fascia.* Although the use of eponyms is discouraged by most writers of anatomy textbooks, there are a few regions of the body, such as this one, where continued clinical usage warrants the medical student's having at least an introduction to the terms. It is important to note

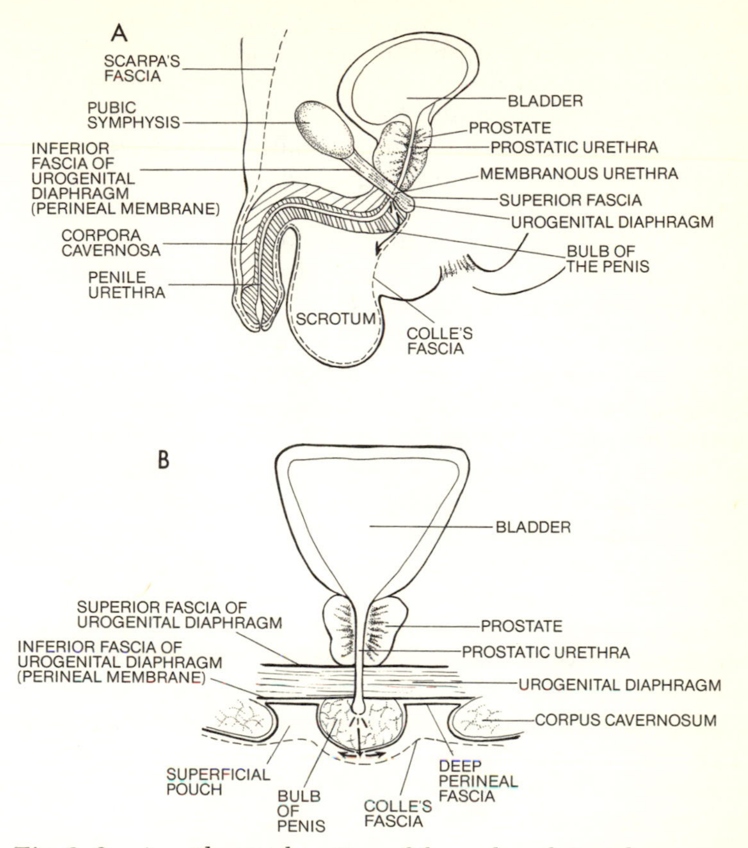

Fig. 6–2. – **A,** midsagittal section of the male pelvis and perineum. The *arrow* shows the course of spread of urine from a ruptured urethra (at the junction of the *membranous* and *penile* portions) into the *scrotum,* penis and inferior aspect of the anterior abdominal wall. **B,** frontal section of the male perineum near the posterior aspect of the *urogenital diaphragm.* As in **A,** the *arrows* indicate the course of extravasated urine.

here that Colle's and Scarpa's fasciae are continuous, simply being named differently in different areas of the body. This layer is attached to the posterior edge of the urogenital diaphragm, to the ischiopubic rami, to the fascia lata of the thigh a short distance (1–2 cm) below the inguinal ligament and to the deep fascia of the anterior abdominal wall at variable distances between the pubic symphysis and umbilical level. The superficial layer, on the other hand, is continuous across these points of attachment with the superficial fascia of the thigh and body wall, and it expands to form the mass of fat in the ischiorectal fossae posterior to the urogenital triangle. The fat of this layer is replaced by smooth muscle fibers in the scrotum and, together with the membranous layer, it forms the *dartos tunic* of that structure.

It can be seen that a closed sac exists between the deep and the superficial perineal fasciae, and this is known as the *superficial perineal pouch* (space, compartment). The rupture in the boy's urethra broke through the deep perineal fascia and thus allowed extravasated urine and blood to fill this pouch. This clinical condition is the most dramatic way to demonstrate the boundaries of the superficial pouch, which are delineated by the fusion of the membranous layer of the superficial fascia to the deep fascia in various places (e.g., posterior border of the urogenital diaphragm, ischiopubic rami). If the student follows the extent of the membranous layer of the superficial fascia and the deep fascia, regardless of terminology, he will readily appreciate how the extravasated fluid will fill the scrotum and penis between these two layers and extend to the lower portion of the anterior abdominal wall, but *not* beyond the places where the two layers are fused. It should be pointed out here that some anatomy textbooks (e.g., that of Gray[1]) define the superficial perineal pouch as simply the narrow space between the inferior fascia of the urogenital diaphragm and the deep perineal fascia. The space between the latter layer and Colle's fascia is referred to in that book as the superficial perineal fascial cleft.

Surgical repair is effected by catheterizing the urethra, dissecting in the midline of the urogenital triangle and suturing the tear in the urethra as well as repairing any associated structures that may have been injured.

CASE 6–3: "LUMPY" KIDNEYS

History

A 39-year-old woman complained of diffuse abdominal pain and low backache. Radiologic examination revealed enlarged, "lumpy" kidneys bilaterally. She was diagnosed as having *congenital polycystic disease,* which upon further examination was found to involve the liver and pancreas also. She was treated surgically by incision and drainage of the renal cysts, but within a year there was evidence of new cyst formation. As a result of this finding and the involvement of the other organs, her prognosis was considered poor.

Questions

Describe very briefly the structure of the kidney.

In which portion of the kidney would the cysts be found?

What might be one explanation of cyst formation based on the embryology of epithelial organs in general?

Discussion

The *kidneys* are reddish brown, bean-shaped organs that present a smooth rounded contour and a medial indentation, the *hilus,* which is the point of entry or exit of the numerous functionally related structures (blood vessels, nerves, lymphatics, ureter). The gross internal structure of the kidney is best observed by sectioning the organ in a frontal plane into anterior and posterior halves (Fig. 6–3). The kidney is bounded by a collagenous *capsule.* Investigation of the cut surface reveals an outer band of pale *cortex* that surrounds an inner *medulla.* Grossly the medulla may be subdivided into a number of *pyr-*

Fig. 6–3. – Frontal section of the kidney.

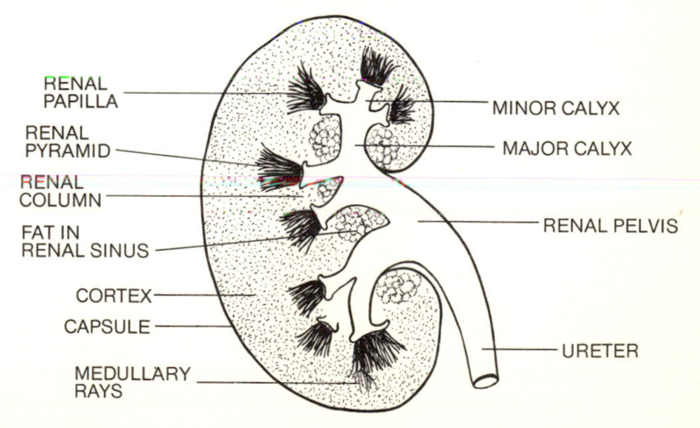

amids, the tips of which drain into a *minor calyx* of the ureteral tree (see also Case 6–4). Cortical tissue extends down into the medulla between the pyramids, where it is referred to as the *renal columns.* At the base of each pyramid are found striations, which extend into the cortex; these are known as *medullary rays.*

The functional unit of the kidney is the *nephron* or *renal tubule,* which is derived from the metanephros. There are more than a million of these units in each kidney. Each consists of a blind end, the *glomerular* (or *Bowman's*) *capsule,* which is invaginated by a tuft of capillaries, the *glomerulus.* The capsule with its tuft of capillaries is collectively known as a *renal corpuscle.* Proceeding distally, the capsule gives way, in order, to a *neck,* a *proximal convoluted tubule* and a straight portion, the descending limb of the *loop of Henle.* It then turns as a hairpin back toward the renal corpuscle as the ascending limb of the loop of Henle, this in turn becoming the *distal convoluted tubule.* The latter then empties into a *collecting tubule,* which is the excretory duct. This terminates on the *papilla* or apex of the pyramid, which in turn empties the contents into the minor calyx. The renal corpuscles, proximal and distal convoluted tubules and proximal portions of the collecting tubules are found in the cortex of the kidney. The loops of Henle and the remainder of the collecting tubules occupy the medulla.

Durham[2] discusses cyst formation from the point of view of the failure of tubules in epithelial organs to unite with the excretory or collecting ducts (a theory subscribed to by numerous embryologists). In this hypothesis, it is assumed that the secretory cells continue to manufacture and discharge their products, the latter accumulating in situ for lack of an excretory mechanism. Thus, the organ undergoes cyst formation and perhaps localized degeneration. By this explanation, we would expect to find the great majority, if not all, of the renal cysts in the cortex, i.e., that portion of the kidney that contains the renal corpuscles.

CASE 6–4: RENAL STASIS AND CALCULUS FORMATION

History

A 24-year-old male paraplegic entered the hospital with excruciating abdominal pain, mainly on the right side. A history revealed that he had sustained a severe injury to the lower spinal cord in an automobile accident at the age of 13. From that time on he had been confined to a wheelchair. The medical examinations following the accident revealed a slowly progressing renal stasis and radiologic evidence of renal calculus formation. The immediate problem was treated as a surgical emergency, and recovery was good. With time, however, there was evidence of additional calculus formation.

Questions

What specific situation was responsible for this patient's pain? What is this condition called?

Briefly describe the anatomy of the ureteral tree from its origin in the kidney to its termination in the bladder.

Discussion

This patient was suffering from *renal colic,* acute pain caused by a kidney stone being lodged somewhere along the course of the ureter, in this case on the right side. The pain, caused by distention of the ureter by the calculus, is perceived in the hypogastric or lumbar regions of the abdomen. If the stone lodges near the entry of the ureter into the bladder, the pain may be referred to the external genitalia. The student might find it helpful to study Case 9–2 also at this time; this deals with the formation and passage of renal calculi caused by different circumstances.

The hilus of the kidney (see Case 6–3) leads into the *renal sinus,* a space lined by a continuation of the capsule into the interior of the kidney. It contains the *renal pelvis,* which is the proximal expanded portion of the *ureter.* Several *minor calyces* combine to form two or three *major calyces* (see Fig. 6–3), the latter of which combine to form the renal pelvis. This in turn narrows in diameter to become the ureter proper, which, like the renal pelvis, occupies a position posterior to the renal vein and artery.

The ureters are muscular tubes about 12 inches in length that extend from the renal pelves on each side to the posterolateral angles of the bladder. About half of their length is abdominal and half pelvic in position. They are retroperitoneal, as are the kidneys. Urine is moved down the ureters in peristaltic waves of smooth muscle contraction, and it enters the bladder in intermittent spurts. The ureters are supplied by branches of the renal, gonadal, inferior vesical and other arteries. Nerve supply is from the renal and hypogastric plexuses of the autonomic nervous system.

REFERENCES

1. Gray, H.: *Anatomy of the Human Body,* Goss, C. M., (ed.) (29th American ed.; Philadelphia: Lea & Febiger, 1973).
2. Durham, R. H.: *Encyclopedia of Medical Syndromes* (New York: Paul B. Hoeber, Inc., 1960).

Chapter 7

HEAD, NECK AND SPECIAL SENSES

CASE 7–1: TUMOR IN THE NECK

History

A retired laborer, age 67, presented with a demonstrable lump in the anteroinferior angle of the posterior triangle of the neck on the right side (i.e., in the angle formed by the posterior border of the sternocleidomastoid muscle and the superior border of the clavicle). The lump, diagnosed as a malignant tumor, extended superiorly above the posterior belly of the omohyoid muscle, i.e., it exceeded the limits of the subclavian or supraclavicular triangle, a subdivision of the posterior triangle of the neck. He complained of some pain in the general area of the tumor and a general reduction in ability to perform certain normal movements of the right upper extremity (e.g., inability to abduct the limb), accompanied by anesthesia or hypoesthesia over parts of the limb. In addition, he had noticed increasing difficulty in finding a pulse in the right wrist and, when he was able to locate it, observed that it was of a lesser magnitude than that of the left upper limb. Neurologic examination revealed paralysis of the supraspinatus and infraspinatus muscles and reduced motor and sensory function of the following nerves: median, musculocutaneous, radial, axillary, lateral pectoral and subscapular. The ulnar, medial pectoral, thoracodorsal, medial brachial cutaneous and medial antebrachial cutaneous nerves were unaffected. Angiograms demonstrated generalized reduction in arterial blood supply to the pectoral area, the shoulder and the entire upper limb. There was no loss of blood supply to the head or neck.

Questions

Identify as specifically as possible the portions of the *two* structures that were compressed by the tumor in order to produce these symptoms.

What types of physical injuries also might be responsible for the neurologic disorders observed in this patient?

Discussion

Besides causing pain locally, the tumor was found to be compressing (1) the *upper trunk of the brachial plexus,* and (2) the *subclavian artery* just lateral to the point where they emerge from under cover of the scalenus anterior muscle (Fig. 7–1).

The upper trunk of the brachial plexus is comprised of the anterior primary rami of C_5 and C_6. It gives origin to the *nerve to the subclavius* and the *suprascapular nerve* before dividing into anterior and posterior branches. Both of these nerves would be implicated in the compression of this trunk, as would any of the other nerves that contain these fibers, e.g., the axillary nerve. Damage to the nerve to the subclavius is difficult to test neurologically (it supplies the subclavius muscle with its limited function and the sternoclavicular joint); damage to the suprascapular nerve, however, would result in denervation of the supraspinatus and infraspinatus muscles, an easily detectable loss. For example, since these muscles

Fig. 7–1.—Relationship of nerves, arteries, veins and muscles to the right *first rib*. The patient's tumor exerted pressure in the angle between the lateral margin of the *scalenus anterior muscle* and the *subclavian artery*.

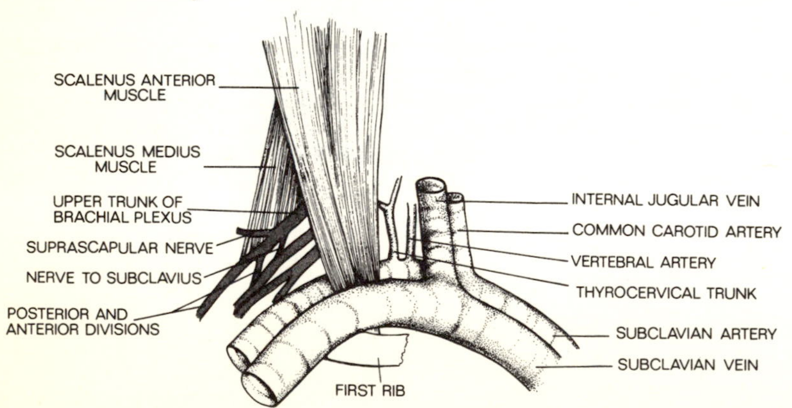

(principally the infraspinatus assisted by the teres minor) rotate the arm laterally, their denervation would produce an abnormal medial rotation of the upper limb due to the unopposed actions of the medial rotators. This symptom is characteristic of "upper type" brachial plexus lesions, and its manifestation is frequently referred to as "waiter's tip" hand (or arm). This type of lesion is most frequently produced by an abnormal widening of the angle between the neck and the shoulder, for instance in a football injury where the upper limb is pulled downward and at the same time the head is forced laterally in the other direction. Such an action puts the upper trunk on the stretch, and serious nerve damage can occur.

When this type of injury occurs during birth, the resultant condition is referred to as *Erb's*, or *obstetric, palsy.* In addition to paralysis of the lateral rotators, the supraspinatus also is denervated or weakened in upper trunk injuries. This condition hampers, but does not eliminate, abduction of the arm. The supraspinatus acts as an initiator of abduction due to its angle of insertion on the humerus. The deltoid, acting alone, would tend to drive the head of the humerus up under the acromion, since its fibers, in effect, run vertically from that point to the deltoid tuberosity on the humerus. Therefore, a patient with supraspinatus paralysis by necessity leans to the side in order to initiate abduction of the arm to the point where the deltoid muscle can take over and complete the action.

"Lower type" brachial plexus injuries may occur when the arm is stretched upward to an abnormal degree. This results in damage to the lower trunk (C_8 and T_1), which supplies most of the intrinsic muscles of the hand. Denervation of these muscles results in a clinical condition known as clawhand (for another way that this symptom can be produced and for a further explanation of the condition, see Case 2–6).

The *long thoracic nerve* (C_{5-7}) and the *dorsal scapular nerve* (mainly C_5) arise from anterior primary rami proximal to the formation of the trunks of the brachial plexus and, therefore, would not be expected to be implicated in this case. Thus, we would not expect to observe the clinical signs characteristic of loss of activity in the serratus anterior (winged scapula) or rhomboid muscles (reduced ability to adduct or retract the scapulae). Just distal to the points of origin of the nerve to the subclavius and the suprascapular nerve, the upper trunk divides into anterior and posterior branches that contribute to the formation of the lateral and posterior cords, respectively. The nerves involved with this example of brachial plexus damage would, therefore, be those that carried the C_5 and C_6 components upon branching and recombination of these cords. The beginning medical student does not need to memorize the components of the individual branches of the brachial plexus; a general understanding of the distribution of the trunks and cords, however, is important. Such knowledge would allow the student to quite accurately pinpoint the site of nerve compression in this particular case.

The subclavian artery was compressed, but not totally occluded, at the point where it crossed the first rib, i.e., just proximal to the point at which the vessel changes its name to the axillary artery. This area of the subclavian artery is defined as its third part. Since all branches of the subclavian artery to the head and neck originate proximal to the lateral border of the scalenus anterior muscle (from the first and second parts of the artery), blood supply to these areas would be unaffected. On the other hand, all of the structures in the upper limb supplied by branches from either the axillary artery or its continuation in the arm, the brachial artery, would be rendered variably ischemic by the reduced blood supply caused by the compression of the subclavian artery. The student might find it advisable to study the anastomoses between branches of the subclavian artery and branches of the axillary artery that would tend to establish collateral circulation around the site of partial blockage (see also Case 4–1).

CASE 7–2: TUMOR OF THE LOWER LIP

History

A 54-year-old college professor had been bothered for a number of months by a small nodule in the central portion of the lower lip. Thinking that this might be related to his heavy pipe smoking of the past 30 years, he gave up the habit, but after 3 months of abstention the nodule had increased in size, prompting him to consult his physician.

The doctor examined the lump visually and then began a systematic palpation of the professor's lymphatic chain from the tip of the chin to the inferior cervical area at the level of the clavicle. He found enlargement of the lymph nodes at the floor of the mouth near the tip of the chin, in the area of the submandibular gland on both sides, and also bilateral involvement of the deep cervical lymph nodes along the internal jugular veins from the posterior belly of the digastric muscle superiorly to just above the omohyoid muscle inferiorly. Needle biopsies revealed a malignant tumor of the lip accompanied by metastatic spread bilaterally to both superficial and deep cervical lymph nodes. The physician was encouraged that this process had not continued much farther than the central portion of the neck. The patient was immediately scheduled for a bilateral radical neck dissection designed to remove the tumor, the involved lymphatic vessels and nodes and all adjacent structures (e.g., muscles, nerves, arteries, veins, glands) that might have been invaded by metastatic tumor cells.

This procedure is a very dramatic one, but in many instances it is totally successful and less debilitating than might be expected. This patient recovered from the operations, was able to return to his position and was free from any detectable cancer 5 years after surgery.

Questions

The student should give considerable thought to which structures may be removed and to the attendant level of disability rendered by their removal. Equally important is an understanding of the relationships of these structures to those that should not be removed or injured, e.g., the position of the internal jugular vein to the common (or internal) carotid artery and vagus nerve.

Trace the course of spread of the cancer to the various named lymph nodes of the head and neck.

In very early detection of cancer of the central area of the lower lip, would it be possible to remove only one set of lymph nodes, thus avoiding radical neck dissection?

Discussion

In cancers occurring in the general area of the mouth (anterior oral cavity), those of the lips are the least malignant and those of the tongue the most malignant. Cancer of the tongue has a tendency to spread bilaterally into the deep cervical lymphatics (with or without first passing through one or more sets of superficial cervical lymph nodes). On the other hand, cancer from the lower lip almost always spreads initially to *superficial lymph nodes* in one of two ways: that from the lateral portions of the lip drains to *submandibular nodes* whereas that from the central portion of the lip drains initially to *submental nodes* and from there to submandibular nodes (Fig. 7–2). This situation with regard to metastases from the central portion of the lower lip is significant since, in very early detection of the cancer, it might be arrestable by removal of the submental nodes only. On the other hand, once the cancer has spread from the submental nodes, it is likely to enter the submandibular nodes bilaterally, thus allowing possible metastasis to both sets of deep cervical lymphatics, as occurred in this patient.

The submental nodes are situated between the anterior bellies of the two digastric muscles at the floor of the mouth. Their efferent vessels, in addition to draining to the submandibular nodes, pass partly to a deep cervical node on the inter-

Fig. 7–2.—Lymphatic drainage from the *central* portion of the *lower lip*. Note that numerous channels exist for the spread of metastatic cells from the superficial lymph nodes to the *deep cervical lymphatic chain*, the latter located along the *internal jugular vein*.

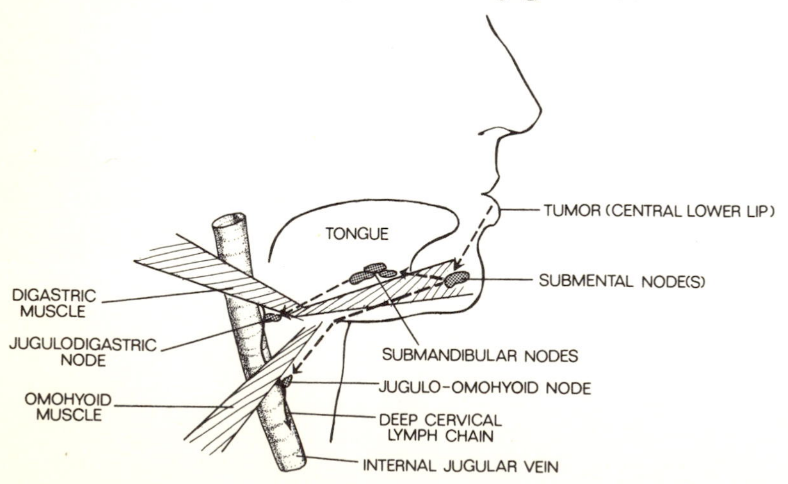

nal jugular vein at the level of the cricoid cartilage. The submandibular nodes lie on the submandibular gland deep to the body of the mandible in the submandibular triangle of the neck. Their efferent vessels drain directly to the superior set of deep cervical nodes. One node of this series, the *jugulodigastric,* which lies in the angle formed by the internal jugular vein and the posterior belly of the digastric muscle, is readily palpable in most individuals. Whereas enlargement of this node may occur in metastatic or infectious spread from the lips or anterior tongue, it more often is diagnostic of disease in the palatine tonsil or posterior tongue. The student should review at this point the lymphatic drainage of the entire head, from superficial lymphatics to deep cervical lymph nodes and chains, to right and left jugular trunks and finally to the thoracic duct on the left and right main lymphatic duct on the right.

With regard to radical neck dissection, all tissues are commonly removed that lie between the platysma muscle and the deep fascial layers of the neck, that is, the sternomastoid and omohyoid muscles, internal jugular vein, accessory nerve, submandibular salivary gland and all lymph nodes and lymphatic vessels. Generally the carotid arteries (common, external, internal), the vagus and phrenic nerves and the trunks of the brachial plexus are left intact, although certain of them (including the carotid arteries and vagus nerve) may be removed if the situation so warrants. It is apparent that a bilateral neck dissection could not result in bilateral removal of many of these structures, as it would create a situation that would not be compatible with life. The following additional structures also may be sacrificed if the particular case so indicates: part or all of the mandible, the infrahyoid (strap) muscles, one lobe of the thyroid gland, the hypoglossal nerve and the lingual branch of the trigeminal nerve. In such severe cases, it is apparent that considerable disability would result. Martin[1] notes, however, that conditions of this type may still be consistent with a productive life. The following is an excerpt from his article on radical neck dissection: "The president of a large chain store corporation survived 14 years and over 50 operations, most of which were for the excision of recurrent cancers originating in the gum. For the last seven years of his life, he was without lower jaw and lip, tongue, larynx, and internal jugular veins. Two years before his death, one common carotid artery had to be sacrificed. Despite all this, he remained mentally and physically active, interested in family and public affairs. He retained the presidency of his firm and was an ardent fisherman."*

CASE 7–3: EXOPHTHALMIC GOITER

History

A 59-year-old woman presented with the characteristic symptoms of exophthalmic goiter (for a complete list of the eponyms for this condition, e.g., the Basedow syndrome, the Graves disease, consult a standard medical dictionary). The symptoms of this disease result from excessive production of thyrotropic hormone by the anterior pituitary and usually from the presence of a diffuse toxic goiter. The predominant and obvious physical symptoms in this patient were a clinically demonstrable goiter (enlarged pulsating thyroid gland), mainly on the left side, and bilateral exophthalmos (abnormally protruding eyeballs). Physical examination and laboratory tests revealed the following additional symptoms: tachycardia (abnormally rapid heart rate, usually considered to be greater than 100 beats per minute), anxiety, profuse sweating (hyperhidrosis), insomnia, recent weight loss, fine muscle tremors and increased basal metabolic rate. The degree of thyrotoxicosis in this patient as determined by protein-bound iodine tests was classified as severe. Therefore, a subtotal thyroidectomy, in which the entire left lobe of the gland was excised, was the therapy of choice, combined with the use of antithyroid drugs and radioactive iodine (^{131}I).

The disease process was abated in the remaining portions of the gland (right lobe and isthmus), and the patient began to recover from her symptoms within a few days. However, new and different symptoms began to appear shortly after surgery. These were manifested as hoarseness, muscular weakness, slight fatigue, occasional cramps in the extremities and tetany. The hoarseness proved to be resistant to treatment, but the other symptoms were alleviated by intake of supplemental dietary calcium and vitamin D. These substances had to be taken indefinitely, however, or the latter symptoms tended to reappear.

Questions

Review the normal anatomy and relationships of the thyroid gland.

Identify the various sources of arterial blood and discuss the various anastomoses around the gland.

Are veins prominent also? Where do they drain?

Describe briefly any important variations in the gland or its blood supply that may be apparent.

What accounted for the hoarseness observed postoperatively in this patient?

What caused the other postoperative symptoms, and why was continued dietary supplementation necessary thereafter?

Would you expect to observe additional symptoms if denervation had occurred bilaterally?

Discussion

The *thyroid gland* (Fig. 7–3), the largest endocrine gland in the body, is composed of two *lateral lobes* connected by an intervening *isthmus.* Each lateral lobe presents an apex that projects superiorly and a base that projects inferiorly. The lateral lobes are related anterolaterally to skin and muscle (sternothyroid and sternohyoid) and medially to the larynx and trachea. They generally extend between C. V. 5 to C. V. 7; the isthmus covers tracheal rings 2–4 anteriorly. The gland possesses a capsule, and external to this it is enveloped by the pretracheal layer of the deep cervical fascia. A *pyramidal lobe,* frequently absent, projects superiorly and to the left from the isthmus. It is attached by fibrous tissue to the hyoid bone; when this tissue possesses muscular fibers, as is sometimes the case, it is known as the *levator glandulae thyroideae muscle.* Accessory thyroid tissue may occur anywhere along the course of the embryonic *thyroglossal duct,* which originally connected the thyroid gland with the pharynx (foramen cecum of the tongue), elsewhere in the neck or even in the mediastinum.

Being endocrine in function, the thyroid gland has a profuse blood supply. This consists basically of bilateral *superior* and

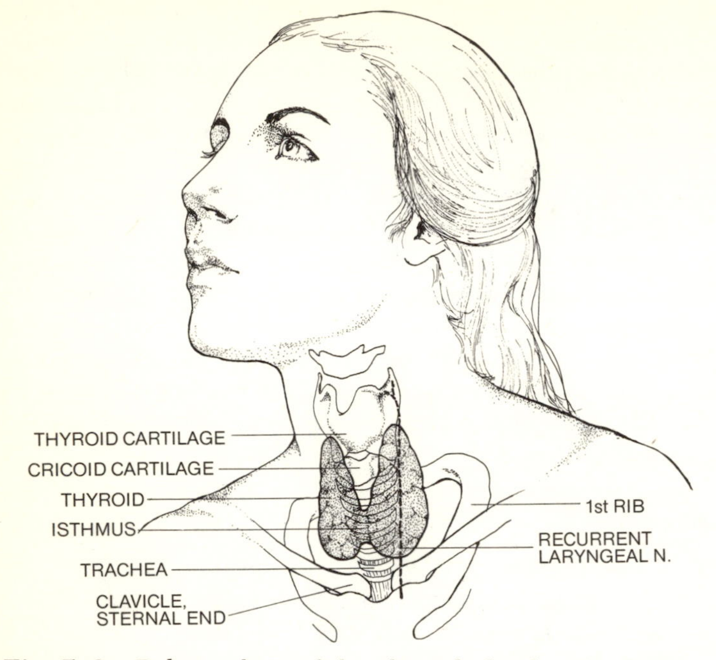

Fig. 7–3.—Relationships of the *thyroid gland* to the larynx and tracheal rings. Note that the *recurrent laryngeal nerve* ascends posterior to the gland and is vulnerable to trauma in thyroid operations.

inferior thyroid arteries. The superior thyroid artery is the second branch to arise from the external carotid artery in the neck just after the bifurcation of the common carotid. It descends in the neck, giving branches to the sternomastoid muscle and to the larynx (superior laryngeal branches), passes deep to the omohyoid muscle (anterior or superior belly) and then enters the pretracheal fascia to ramify around the apex of the lateral lobe. The inferior thyroid artery is a branch of the thyrocervical trunk of the subclavian. It follows a superomedial course, dividing into numerous branches that pierce the fascia independently to form a network around the base of each lateral lobe. Anastomoses occur between the superior and inferior thyroid arteries of each side (vertical anastomoses) and between the two superior arteries and the two inferior arteries above and below the isthmus, respectively (superior and inferior horizontal anastomoses). These connections are profuse and must be kept in mind during surgical intervention in this area. An additional artery, the single *thyroidea ima,* which occurs infrequently, may be found ascending on the trachea in the midline of the neck toward the isthmus and lower poles of the thyroid gland. This artery may originate from the arch of the aorta, the brachiocephalic trunk or occasionally from the right common carotid artery. It is of surgical importance in that, if present, it may be incised in emergency tracheostomy procedures.

The thyroid veins are equally plentiful as the arteries. They develop from a plexus of veins that occupies a position on the surface of the gland. The *superior thyroid veins* coalesce near the apex of each lateral lobe and drain to the internal jugular vein on each side. The *middle thyroid vein* of each side drains the middle portion of the gland and similarly empties into the internal jugular. The *inferior thyroid veins,* which drain the inferior poles of the lateral lobes, descend to join the brachiocephalic veins (usually the left) in the root of the neck.

The *parathyroid glands,* also endocrine, number between one and three on each side. They occupy a position on the posterior surface of each lateral lobe of the thyroid gland (hence their name) between the capsule and fascial sheath of the gland. They are often described as being of the approximate size and shape of a pea (about 5 mm in diameter). The surgeon must be completely familiar with the relationship of these small structures to the thyroid gland; otherwise they might be inadvertently removed along with a lobe of the thyroid, as occurred in this particular case. The parathyroid glands are concerned with calcium metabolism, maintaining a balance between bone and blood. Removal of the parathyroids results in a decrease in blood calcium, which in turn leads to muscular spasms and weakness and nervous hyperexcitability. This condition, which is called *tetany,* will result in death unless dietary intervention (i.e., dietary calcium and vitamin D and/or parathyroid extract) occurs. In this patient only half of the ordinary parathyroid complement was accidentally removed, but this was sufficient to produce the symptoms listed in the case history. Additional symptoms that would be expected to

"Spiker and Sponge."

"Who or what is missing in *The Phantom Tollbooth*?"

"Rhyme and Reason."

The librarian nodded. Even at her young age, it was hard for Monica not to nod along. Rhyme and reason were so clearly far away. Were so clearly not friends of her father's.

Every Wednesday, Monica sat on the library steps and waited for her father's red Datsun to pull up to the curb. In the autumn, when the thunderstorms came, he'd park the little red station wagon in the driveway instead of the garage, and let her roll out her sleeping bag in the back. She'd open the rear window and read her books as the aroma of tire-crushed worms and ozone melded with the remnants of her father's cologne, as the rain spoke to her in Morse code on the car's metal frame. As she turned each page, she'd glance at the sky and know in her bones that this was as close as it got to happiness.

The late afternoon sun began to fade and the spittle of rain clouds defaced the covers of Monica's books. She gathered them from around her on the library steps and stuffed as many as she could into her canvas grocery bag. Her mother's beige Corolla pulled up to the curb. Monica didn't move. Her mother leaned over the passenger seat and let the door swing open above the sidewalk. Monica gathered the remaining books into a stack in her arms, hugging them tight as she descended the library stairs.

"We can't keep doing this," her mother said as Monica slumped into the passenger seat. "You're supposed to get on the bus."

Monica didn't respond.

Her mother stared at her. "Will you at least shut the door, please?" Monica held the stack of books to her chest, afraid of what might happen if she loosened her grip.

Her mother jumped from her seat. She stomped around the front of the car and stood on the sidewalk with her hands on her hips. She grabbed the books from Monica's lap and threw them into the back seat. They tumbled across the torn fabric and onto the muddy floor—the princesses and the frogs, the horses with rainbow manes, the life lessons and happy endings. Those things, those lives where everything worked out in the end and the father walked the young bride down the be observed in untreated underactivity or parathyroid removal are associated with the normal calcification of bones and teeth. In order to prevent these disorders, dietary supplementation has to be taken on a continuing basis.

The postoperative hoarseness observed in this patient was the result of accidentally cutting the left *recurrent laryngeal nerve* during surgical removal of the left lobe of the thyroid gland. This nerve, which is a branch of the vagus (it is sometimes referred to as the inferior laryngeal nerve), *recurs* around the arch of the aorta and then ascends toward the larynx in the groove between the trachea and esophagus. It is in this position as it passes the thyroid gland (see Fig. 7–3) that it is in jeopardy of being sectioned during surgical procedures in the area. The nerve divides on the posterior aspect of the larynx into branches that supply all of the muscles of this structure with the exception of the cricothyroideus. The *posterior cricoarytenoid muscle* is the only abductor of the vocal folds (true vocal cords). Denervation of this muscle by cutting one recurrent layrngeal nerve results in unilateral adduction of the respective vocal fold (due to the now unopposed action of the cricothyroideus, which acts as an adductor by stretching the vocal fold). The demonstrable physical result in this case is hoarseness, since one vocal fold is still "working" and one is not. This condition also may be the result of advanced syphilis, in which case aortic aneurysms (which may impinge on the left recurrent laryngeal nerve) are frequent. The student should recognize that *bilateral* section of the nerves to both posterior cricoarytenoid muscles would result in immediate closure of the *rima glottidis,* which without immediate intervention is incompatible with life.

CASE 7–4: GUNSHOT WOUND IN THE INFRATEMPORAL FOSSA

History

A hunter was accidentally shot from long range with a standard low powered .22 caliber rifle. The bullet, which was relatively spent on impact, entered the right infratemporal fossa at a point just inferior to the zygomatic arch about 3 cm anterior to the external auditory me-

atus. It passed between the condyle and coronoid process of the mandible, i.e., through the mandibular notch, crossed in a horizontal plane and came to lie against bone at the superomedial angle of the fossa (at the junction of the greater wing of the sphenoid bone with the lateral pterygoid plate of that bone). The patient was taken to the hospital in a conscious state, with minimal bleeding from the site of the wound. The bullet was located radiographically and removed by an uncomplicated surgical procedure. The wound healed cleanly, leaving only an insignificant and slightly noticeable scar. Upon neurologic examination, however, immediately after the accident and in follow-up procedures, it was noticed that the patient was lacking a great deal, if not all, of the normal function of the right side of the jaw. Tests revealed a lack of tension in the temporomandibular joint and greatly reduced or absent electromyographic potentials in the medial and lateral pterygoid muscles, the temporalis and the masseter. There was sensory loss in the skin and mucous membrane of the cheek, the lower lip and chin, gums, a portion of the auricle and the temporal area, anterior two thirds of the tongue, and lower teeth of the entire right side. In addition, the mylohyoid muscle and the anterior belly of the digastric were paralyzed on that side, and secretion from the parotid gland was greatly reduced. Interestingly, there was no loss of taste sensation from the anterior two thirds of the right side of the tongue, and secretory activity in the right submandibular and sublingual glands was normal.

Questions

What major structure was severed by the bullet?

How do you account for the lack of interference with taste?

Why was one salivary gland affected but not the other two on the injured side?

Would you expect any major vascular damage to be caused by the bullet?

Discussion

The bullet had enough velocity to sever the *mandibular trunk of the trigeminal nerve* (cranial nerve V, subdivision 3; V_3) but not to penetrate the sphenoid bone, which forms the superior and medial boundaries of the infratemporal fossa. The trigeminal nerve is composed of a sensory and a motor root. The trigeminal or semilunar ganglion occupies a shallow impression in the middle cranial fossa at the side of the dorsum sellae of the sella turcica overlying the foramen lacerum.

It expands laterally onto the anterior slope of the petrous portion of the temporal bone, which forms the posterior limit of the middle cranial fossa. Three nerves reach the ganglion from peripheral areas of the head via named foramina of the skull: the *ophthalmic nerve* (V_1), which traverses the superior orbital fissure; the *maxillary nerve* (V_2), which traverses the foramen rotundum; and the *mandibular nerve* (V_3), which traverses the foramen ovale. The ophthalmic and maxillary nerves are sensory (general somatic afferent) from various parts of the head, e.g., orbit, face, nasal cavity. The mandibular nerve is both sensory (general somatic afferent) from the general area of the infratemporal fossa and lower jaw (see below) and motor (special visceral efferent via fibers of the motor root) to the muscles of mastication.

The foramen ovale allows communication between the middle cranial fossa and the infratemporal fossa and transmits, in addition to the mandibular nerve, the accessory meningeal artery, when present, and sometimes the lesser petrosal nerve. Immediately inferior to the opening of the foramen ovale into the infratemporal fossa, the trunk of the mandibular nerve gives off a *meningeal branch,* which returns to the cranial cavity via the foramen spinosum, and the *nerve to the medial pterygoid muscle.* The trunk then separates into *anterior* and *posterior divisions* (Fig. 7–4). The anterior division, with the exception of the *buccal nerve,* which is sensory to the skin and mucous membrane of the cheek and to the gums, is composed of motor fibers to the muscles of mastication *(masseteric nerve, deep temporal nerves* and *nerve to the lateral pterygoid).* The posterior division, with the exception of motor fibers carried to the mylohyoid and digastric (anterior belly) muscles in the *mylohyoid branch of the inferior alveolar nerve,* is sensory. Its branches are the *auriculotemporal nerve* to the external ear and temporal area; the *lingual nerve,* which supplies the anterior two thirds of the tongue with general somatic afferent fibers; and the *inferior alveolar nerve* to the lower teeth and gums and, by way of its *mental branch,* to the skin of the chin and lower lip.

In addition to denervating all of the structures supplied by the nerves described above, the bullet also destroyed the *otic*

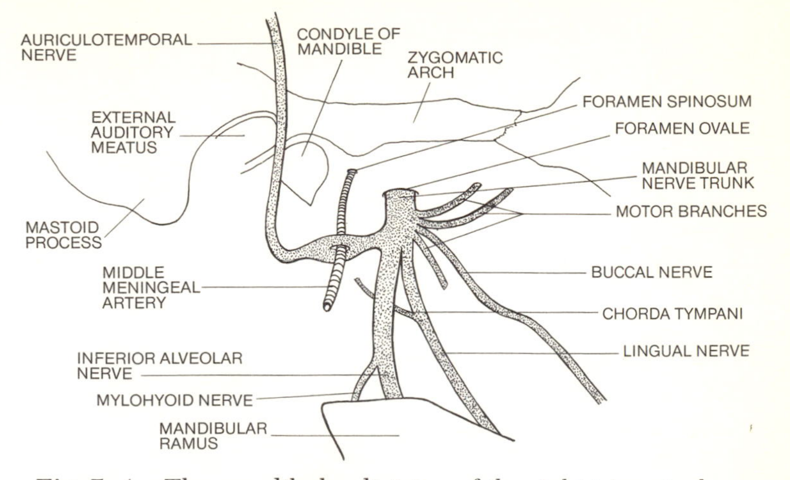

Fig. 7–4.—The mandibular division of the right trigeminal nerve (cranial nerve V) in the infratemporal fossa.

ganglion, which lies just below the foramen ovale immediately medial to the trunk of the mandibular nerve. The roots of the otic ganglion are as follows: (1) the *lesser petrosal nerve,* a branch of the glossopharyngeal (cranial nerve IX), which carries preganglionic parasympathetic fibers to the ganglion; and (2) a *sympathetic root* that carries postganglionic fibers from the superior cervical ganglion to the otic ganglion via a plexus on the middle meningeal artery. The preganglionic parasympathetic fibers synapse with postganglionics within the otic ganglion; the latter fibers along with the sympathetic fibers then reach the parotid gland via the auriculotemporal nerve. The parasympathetic fibers are secretomotor to the parotid gland; the sympathetic fibers are vasomotor to the blood vessels of that gland.

The symptoms of our patient can, therefore, be explained by a lesion of the mandibular nerve and destruction of the otic ganglion. The observation that taste sensation from the anterior two thirds of the tongue was not impaired is explained by the fact that the *chorda tympani,* which carries the special visceral afferent fibers to the brain, leaves the lingual nerve prior (distal) to the site of the lesion. Figure 7–4 shows that the chorda tympani crosses the infratemporal fossa from the petrotympanic fissure to the lingual nerve at a considerably lower level than that of the lesion. The chorda tympani also carries preganglionic parasympathetic fibers (general visceral efferents) that join the lingual nerve to eventually synapse with postganglionics in the submandibular ganglion. These latter fibers rejoin the lingual nerve to be distributed to the submandibular and sublingual glands, where they are secretomotor. Thus in the patient, function of the parotid gland was impaired, whereas that of the submandibular and sublingual glands was not affected. Very little bleeding was encountered, since the lesion was at a higher level than that occupied by either the maxillary artery or the pterygoid plexus of veins. The accessory meningeal artery, even if present and severed, is too small to create any significant problem.

CASE 7–5: TUMOR OF THE PAROTID GLAND

History

A 48-year-old male patient was diagnosed as having a malignant tumor of the right parotid gland necessitating surgical removal. During the operative procedure, the surgeon noted that the tumor extended to the deep lobe of the gland and therefore removed the entire gland. The patient recovered from the operation and was sent home in a few days. He was seen as an outpatient on a regular basis for several weeks of postoperative radiation therapy to eliminate any possible metastases that might have occurred to the superficial and deep cervical lymphatic chains. Five years after the original surgery he was free from any cancer and was pronounced cured. He was not as fortunate as the history might suggest, however, because immediately after the surgery, the following symptoms were observed on the ipsilateral side of the face: (1) The mouth was displaced as a result of unopposed contraction of the muscles on the uninjured side. Thus, the corner of the mouth drooped, and the patient was not able to smile, whistle or blow in the normal fashion. (2) The nasolabial furrow was less pronounced, and the wrinkles on the right side of the forehead had become smoother. (3) He had ptosis of the upper right eyelid, the lower lid was slightly everted and he could not close the right eye or blink. As a result, he developed conjunctivitis and eventual corneal scarring, which led to blindness in the right eye. Further testing revealed that there was no decreased lacrimation, the patient was not hypersensitive to sounds, there was no loss of taste sensation from the anterior two thirds of the tongue and maximal depression of the lower jaw did not cause deviation of the tongue or mandible to the unaffected side.

Questions

Based on the information presented in this case, state what structure was incised during the operation for removal of the patient's diseased parotid gland.

Review the overall anatomy of this structure both centrally and peripherally, especially its relationship to the parotid gland.

Discuss the specific losses observed with regard to your knowledge about the distribution of this structure.

Why was there severe loss of muscular function, but no impairment of lacrimation, taste or hearing?

What might be some other ways that this condition could be brought about?

Is the condition always permanent, as in this patient, or can it sometimes be transient?

Discussion

The injured structure was the right *facial nerve,* the main trunk of which was incised during surgical removal of the parotid gland. The injury resulted in a condition known as *Bell's palsy* or *peripheral facial paralysis.*

The facial nerve has two subdivisions, a larger one that supplies motor fibers to the muscles of facial expression, and a smaller one, the *nervus intermedius,* that contains fibers carrying taste sensation from the anterior two thirds of the tongue and secretomotor impulses to the lacrimal, submandibular and sublingual glands (Fig. 7–5). The combined subdivisions of the nerve enter the facial canal in the temporal bone, through which they run a tortuous course. The nerve bends sharply at the *geniculate ganglion* (which contains the cell bodies of the afferent taste fibers from the anterior two thirds of the tongue), lies in close relation to the middle ear and then exits from the skull via the *stylomastoid foramen.* It then immediately en-

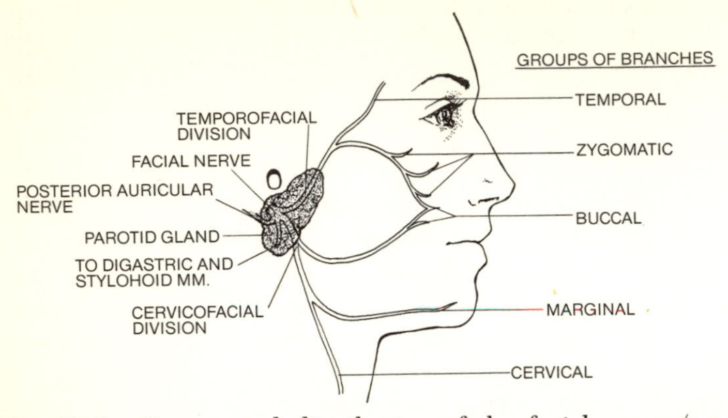

Fig. 7–5.—Course and distribution of the facial nerve (cranial nerve VII) in the face and superior aspect of the neck (based on Gardner *et al.*[2]).

ters the parotid gland, which lies just anterior to this foramen, and breaks up within the gland into its terminal branches to the face.

Within the facial canal the nerve gives rise to the following branches: (1) the *greater petrosal nerve,* which carries parasympathetic secretomotor fibers to the lacrimal and nasal glands, synapsing in the pterygopalatine ganglion; (2) the *nerve to the stapedius,* which supplies that middle ear muscle; and (3) the *chorda tympani,* which carries taste fibers from the tongue and parasympathetic secretomotor fibers to the submandibular, sublingual and lingual glands, the latter fibers synapsing in the submandibular ganglion (see Case 7–4). Immediately after its exit from the stylomastoid foramen, the nerve gives off muscular branches to the stylohyoid and posterior belly of the digastric and the *posterior auricular nerve,* which carries motor fibers to the posterior auricular and occipitalis muscles and sensory fibers to the auricle. The main trunk of the nerve divides within the parotid gland into *temporofacial* and *cervicofacial trunks,* which further divide and recombine to form the *parotid plexus.* The parotid gland is divided into superficial and deep lobes. The facial nerve occupies the isthmus between the two lobes and, therefore, damage is almost unavoidable if any manipulation of the deep lobe is necessary, as it was in this instance. The parotid plexus divides into five named branches, which supply general areas or groups of muscles of the face: (1) *temporal branches* to the anterior and superior auricular muscles, the frontalis and the superior aspect of the orbital area; (2) *zygomatic branches* to the inferior orbital area and the area of the cheek (zygoma); (3) *buccal branches* to the buccinator and other muscles of the mouth; (4) *mandibular branches* to the muscles of the chin and (5) the *cervical branch,* which runs deep to and supplies the platysma muscle. There is a great interchange (anastomosis) among many of these branches and, to a lesser degree, among branches of the facial and trigeminal nerves in the face.

It is apparent from the symptoms in this patient that the facial nerve was cut *after* (distal to) giving off all but the branches to the muscles of facial expression. Thus, there was no loss of secretory activity in any of the glands innervated by the facial nerve (lacrimal, nasal, submandibular, sublingual, lingual), no loss of taste from the anterior two thirds of the tongue (chorda tympani), no hearing problems (nerve to the stapedius), no loss of activity in the posterior auricular, occipitalis, stylohyoid and digastric (posterior belly) muscles and no sensory loss to the auricle. The muscles of facial expression demonstrate a normal resting tone and, therefore, denervation caused by a lesion in the facial nerve results in relaxation of the expressive lines of the face as well as in the more grossly observable losses described in this patient's history. The student is advised to read at this time a standard textbook description of the facial muscles and their actions.

Bell's palsy may be caused in a number of ways, e.g., a blow to the parotid area, toxic substances and cold. It is not uncommon for people who have driven with the car window open or slept by an open window in very cold weather to demonstrate the signs of peripheral facial paralysis. In many of these instances the symptoms are transient since nerve damage is incomplete, and recovery occurs over variable lengths of time. The prognosis in patients in whom the nerve is actually transected, as in this specific instance, are poor, however, unless the injury is simple and the cut ends of the branch or branches can be rejoined immediately.

CASE 7–6: INFECTION OF THE CHEEK

History

A teen-age girl with severe acne was in the habit of squeezing and picking the pustules on her face. As a result of this manipulation, one pimple located just lateral to the nasolabial furrow on the left cheek became infected. Thinking that the pimple was simply getting worse, she continued to squeeze and pick at it from time to time. Within a few days of the onset of the initial infection, the girl complained of a headache and mild, diffuse "pain near my back teeth on the left side." She also observed a faint red line extending subcutaneously from the infected pimple toward the medial angle of her eye. None of these complaints was thought to be associated with the infected pimple, which was still inflamed but did not seem to her to be enough of a problem to consult a physician. However, two days later she was febrile and suffering from a general malaise. She demonstrated signs of very early exophthalmos, edema of the eyelids, edema and congestion of the conjunctiva, a slight ptosis and some impairment of the normal range of movements of the eyeball. An emergency appointment was made with an ophthalmologist who, after a brief examination of the girl, placed her immediately in the hospital. She was placed on massive dosages of antibiotics, but the symptoms were slow to subside. She was, however, taken off the critical list within a week and released from the hospital 10 days later. Her physician kept her on oral antibiotics for 2 weeks after her release from the hospital and pronounced her cured at the end of that time.

Questions

To what major intracranial structure did the infection spread in order to cause the symptoms observed in the left orbit?

Identify the different courses by which the infection was able to spread from the cheek to within the cranial cavity?

How do you explain the slight pain felt by the girl in the general area of the left mandibular molars?

How would you explain the girl's slow response to antibiotic treatment?

Discussion

The infection in the cheek reached the *cavernous sinus* within the middle cranial fossa by traversing various veins of the face, orbit, infratemporal fossa and base of the skull. The cavernous sinus, one of the venous sinuses of the dura mater, lies at the side of the body of the sphenoid bone and extends from the superior orbital fissure anteriorly to the apex of the petrous portion of the temporal bone posteriorly. The space (sinus) formed between the dura and the bone is trabeculated and contains the internal carotid artery in its central portion and cranial nerves III, IV and VI plus the ophthalmic and maxillary branches of the trigeminal nerve along its lateral wall (Fig. 7–6, A). The venous blood that percolates through the intertrabecular spaces of the sinus is separated from the artery and nerves by endothelium. The cavernous sinus communicates with the orbit via the *superior* and *inferior ophthalmic veins*, with the transverse sinus via the *superior petrosal sinus* and with the superior bulb of the internal jugular vein through the *inferior petrosal sinus*. In addition, it drains the *sphenoparietal sinus*, which runs along the free edge of the lesser wing of the sphenoid bone; communicates with the *pterygoid plexus of veins* in the infratemporal fossa by emissary veins that traverse the base of the skull; and is connected with the cavernous sinus of the opposite side via *anterior* and *posterior intercavernous sinuses*. The latter form a circle of veins around the hypophysis known as the *circular sinus*.

The portion of the girl's face that contained the infected pimple has been termed the "danger area" due to the numerous connections of the veins in this vicinity with veins that enter the cranial cavity (Fig. 7–6, B). In addition, these veins do not for the most part possess valves, so that pressure, e.g., squeezing a pimple, can cause infectious material to be propelled in a direction retrograde to the normal flow of blood. One course the infection followed in this case, as evidenced by the reddish line extending from the girl's cheek toward the medial angle of the eye, was in the *angular vein*, the main tributary of the facial vein. The angular vein anastomoses with the superior and inferior ophthalmic veins, which cross the orbit from anterior to posterior and pass through the superior orbital fissure to enter the cavernous sinus.

Another tributary of the facial vein, the *deep facial vein*, communicates with the *pterygoid plexus of veins*, which lies

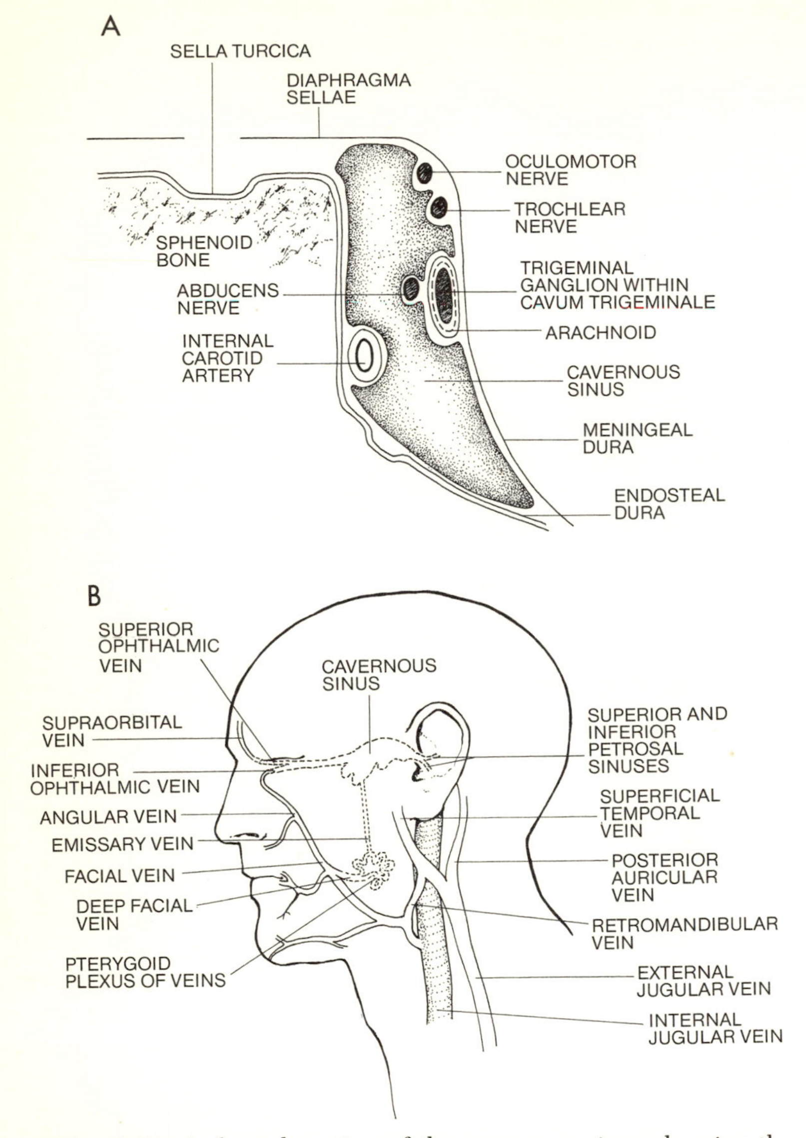

Fig. 7–6.—**A,** frontal section of the cavernous sinus showing the relationships of various nerves and the *internal carotid artery.* **B,** venous drainage of the face showing connections of superficial veins to deeper structures *(pterygoid plexus* and *cavernous sinus).*

within the infratemporal fossa. The manipulation of the infected pimple also forced pus from the cheek into this plexus of veins, which probably explains the diffuse pain felt by the girl in the region of the left posterior lower teeth. The pterygoid plexus of veins also has communications with the cavernous sinus through *sphenoidal emissary veins,* which traverse the floor of the cranial cavity in the foramen ovale, foramen lacerum and foramen of Vesalius. It is possible, therefore, that the infection could spread to the cavernous sinus by one or more of these routes as well. It also could spread to the systemic circulatory system through the connections of the facial vein with the jugular system of veins in the neck and/or to the internal jugular vein via the superior and inferior petrosal sinuses.

Blood flow through the cavernous sinus is sluggish and, as a result, the infection would tend to stagnate there and to increase in size and severity. This condition also explains why the infection demonstrated resistance to the blood-borne antibiotic, and why it was only after repeated massive doses that the infection finally began to subside. Some investigators believe that the normal pulsations of the internal carotid artery assist in the movement of blood through the cavernous sinus. Pressure from the infection on the nerves that occupy the lateral wall of the sinus explains the impairment of the normal movement of the eyeball (cranial nerves III, IV and VI to the extraocular muscles), as well as the slight ptosis of the upper eyelid (cranial nerve III to the levator palpebrae superioris muscle). The restriction of return of venous blood from the orbit produced the exophthalmos and edema of the eyelids and conjunctiva. All of these disorders were alleviated with elimination of the infection.

CASE 7–7: ANEURYSM OF THE INTERNAL CAROTID ARTERY

History

An elderly man demonstrated a developing aneurysm of the right internal carotid artery, as revealed by a bruit heard over the vessel during a routine physical examination. Radiologic examination revealed that the arterial dilatation was located about 4 cm inferior to the base of the skull at the level of C. V. 2. Since the patient was asymptomatic with regard to the aneurysm, the physician decided that, rather than subject him to surgery at that time, he would watch the possible progression of the condition at regular intervals. During his next appointment 3 months later, the doctor observed a slight miosis, ptosis, anhidrosis, and vasodilation of the peripheral arteries (thus causing a flushing of the skin), all symptoms occurring on the right side of the face. Another appointment was set for 1 month thence, at which time the patient demonstrated a progression of the major symptoms described above. In addition, the following secondary symptoms were manifested at this time: an illusion of slight enophthalmos, ocular hypotony, slight facial hemiatrophy (diminished muscular tone on the right side of the face; slight tissue wasting), transitory increases in facial temperature on the right side and increased secretion of tears from the right eye. The physician, recognizing that the symptoms were the indirect result of the aneurysm, scheduled the patient for surgery. An aneurysmectomy (surgical removal of the sac) was performed, the patient recovered quickly and the symptoms of the disorder diminished gradually over a period of several weeks and finally disappeared altogether.

Questions

What structure was compressed by the aneurysm?

Attempt to account for the major symptoms (miosis, ptosis, anhidrosis, vasodilation) on the basis of this compression.

What are the origin and course of the fibers contained within the compressed structure?

What other ways can you think of that this structure or related components might become injured?

How do you account for the disappearance of symptoms after the patient's operation for removal of the aneurysm?

Discussion

This condition is known as the *Horner syndrome* and is the result of interruption of outflow of the sympathetic nerve fibers to the head. In this particular instance the condition was caused by compression of the *superior cervical sympathetic ganglion* by the aneurysm of the internal carotid artery. The *autonomic nervous system* is comprised of two components, the *sympathetic (thoracolumbar) outflow* and the *parasympathetic (craniosacral) outflow*. Both are composed of general visceral efferent fibers. The two components for the most part are antagonistic, serving the function of maintaining a balance of forces acting on the numerous viscera of the body. Thus, the sympathetic division of the autonomic nervous system increases the heart rate, whereas the parasympathetic slows it. In contrast, the sympathetic fibers slow the activity (peristalsis) of the gut, and the parasympathetic system increases the digestive process. The sympathetic system is responsible for the so-called "fight or flight" reaction; in other words, it sets in action those bodily functions that would be required for either fighting or running from an adversary, i.e., an increased heart rate (note that when we even *imagine* danger, sympathetic impulses elicit this sort of response). The parasympathetic division, on the other hand, is responsible for "homeostatic" or vegetative responses, e.g., salivary and digestive gland secretions. Certain teleologic arguments have been put forth to explain various aspects of the autonomic response; for example, a dog's fur stands up when he is frightened (sympathetic outflow) in order to cool his body, therefore aiding him physiologically whether he chooses to fight or run away.

With regard to the structures of the face affected in the Horner syndrome, the sympathetic division of the autonomic nervous system sends motor impulses to (1) the *dilator pupillae muscle* of the eye, which, as its name implies, dilates the pupil upon contraction; (2) the *superior tarsal muscle*, which consists of smooth muscle fibers and aids the levator palpebrae superioris muscle in the support and elevation of the upper eyelid; (3) the *sweat glands* and (4) the smooth muscle of *peripheral blood vessels*. Thus, interruption of the sympathetic outflow to the face leaves the sphincter pupillae muscle (innervated by parasympathetic fibers from cranial nerve III) unopposed, resulting in a constricted pupil (miosis). The other three symptoms are caused simply by the absence of sympathetic innervation to the structures involved, i.e., paralysis of the superior tarsal muscle (ptosis), loss of innervation of the sweat glands (anhidrosis) and relaxation of arteriolar smooth muscle (vasodilation), resulting in redness of the face and increased temperature. Some of the secondary symptoms of this syndrome are more difficult to explain by a direct relationship to sympathetic denervation. The four main symptoms, however, are quite adequate in diagnosing the Horner syndrome and are the ones that should be stressed to the beginning student.

The subject of central control over the autonomic nervous system is complex and will be covered by the student in the various neuroanatomy offerings. Nevertheless, it should be pointed out that lesions in certain tracts emanating from the hypothalamus and traversing the midbrain, pons, medulla and cervical spinal cord will result in the symptoms of the Horner syndrome. However, most of the examples of this disorder probably occur from interruption of the fibers ascending in the cervical sympathetic chain or from damage to the superior cervical ganglion. Cell bodies of preganglionic sympathetic fibers lie in the lateral horn (intermediate gray column) of the upper thoracic spinal cord. Those to the eye and orbit probably originate in the first thoracic segment of the cord and enter the first thoracic nerve, since section of the sympathetic chain below this level does not affect the eye. Preganglionic fibers to the remainder of the face probably originate as far inferiorly as the fourth thoracic cord segment. The preganglionic fibers enter the thoracic sympathetic ganglia through white rami communicantes and then ascend in the paravertebral chain to synapse on cell bodies of postganglionic neurons in the superior cervical ganglion located just beneath the base of the skull. The ganglion extends from C. V. 1 to C. V. 2 (or C. V. 3) and lies between the internal carotid artery and longus capitis muscle. Postganglionic fibers from the ganglion enter a plexus on the external carotid artery as well as forming the *internal carotid nerve,* which ascends on the internal carotid artery to supply intracranial structures. Fibers to the eyeball and orbit are carried on the ophthalmic artery (branch of the internal carotid) and its respective branches. Postganglionic neurons to the face are transported via the facial artery (branch of the external carotid) and by branches of other arteries that supply the face, e.g., the transverse facial branch of the superficial temporal artery and the infraorbital branch of the maxillary artery. The frontal (anterior) branch of the superficial temporal artery also carries postganglionic sympathetic fibers to the anterior portions of the scalp, temple and forehead.

It can be seen from the position of the superior cervical ganglion between the internal carotid artery and the vertebral column (with a small amount of muscle interspersed) that it is very susceptible to compression by an aneurysm or tumor. Whiplash injuries that tend to put the cervical sympathetic chain on the stretch also may cause symptoms characteristic of the Horner syndrome. Such symptoms also may arise from any other trauma or pressure to the cervical and upper thoracic sympathetic chain, e.g., a cervical rib, hemorrhage, goiter and surgical procedures. The patient's symptoms disappeared completely in this case because the pressure was gradual from a slowly developing aneurysm, and the operation was performed at an early stage, thus allowing recovery or regeneration of the injured cell bodies and fibers within the ganglion. In patients in whom the cervical chain was cut or in ischemic accidents to the brain stem or spinal cord, recovery obviously would be incomplete to absent. See Fig. 7–9 for the relationship of the superior cervical ganglion to the internal carotid artery.

CASE 7–8: INFECTION OF THE SCALP

History

A 12-year-old boy in a wrestling match with a friend was scratched on the scalp with a dirty fingernail. The scratch, which bled briefly and then stopped, occurred about two inches posterior to the vertex of the head, 1 inch lateral to the midline on the right side. The scratch was ignored until several days later when an abscess formed at the site. The abscess did not respond to local first aid treatment by the boy's parents, and 2 days later he was taken to the emergency room of a local hospital. At this point he was suffering from a headache, fever and general malaise. The attending physician determined that the symptoms were the result of systemic blood poisoning emanating from the initial scratch and subsequent abscess on the scalp. The boy was immediately placed on antibiotics, and the abscess was drained and cleaned. Response to treatment was rapid, and the boy was released, symptom free, from the hospital a few days later.

Questions

Identify the precise course by which the infection spread from the site of the wound to the systemic circulatory system.

Consider in your diagnosis any alternative courses of spread within the cranial cavity and what conditions might be expected as a result.

What conditions would tend to facilitate the spread of infection from the scalp to the cranial cavity?

Does blood normally flow from the scalp toward the brain or vice versa?

What layer of the scalp is most conducive to the spread of infection? Why?

Discussion

The *scalp* consists of five anatomically recognizable layers, from without inward, as follows: (1) skin, (2) subcutaneous tissue, (3) galea aponeurotica, (4) subaponeurotic tissue and (5) periosteum (pericranium). The subcutaneous tissue (layer 2) contains blood vessels and nerves that supply the overlying skin, and emissary veins are prominent in the loose subaponeurotic tissue (layer 4). In this case, the bleeding from the scratch undoubtedly resulted from extension of the simple wound only into layer 2 of the scalp. Had the scratch been cleaned immediately, no further complications would have been expected. The fact that it was not, however, allowed subsequent formation of the abscess, which eroded the underlying aponeurosis and was thus free to spread within the *emissary veins* in layer 4. This subaponeurotic layer has been termed a "danger area" because of the ease of spread of infection within it and, more importantly, because the infection can enter the cranial cavity via these veins. It is likely that the parents' manipulation of the abscess facilitated the spread to intracranial structures.

Examination of the human skull commonly reveals a small *parietal emissary foramen* on each side of the midline; in this patient the site of the abscess was very near this foramen on the right side. From the emissary vein, which traverses the foramen, the infection spread to the *superior sagittal venous sinus* (Fig. 7–7), which lies within the *falx cerebri,* the dural reflection extending into the longitudinal fissure between the two cerebral hemispheres. The superior sagittal sinus lies just beneath the internal surface of the calvaria, beginning anteriorly at the crista galli of the ethmoid bone. It receives *superior cerebral veins* and *lateral lacunae,* the latter of which receive *meningeal veins* and *diploic veins.* The diploic veins lie within the diploë of the skull between the two tables of compact bone; they also communicate with the veins of the scalp and may allow infection to spread to intracranial structures. Posteriorly, near the internal occipital protuberance, the superior sagittal sinus either joins the right transverse sinus or enters into the *confluence of the sinuses* with the *straight sinus,* the *right* and *left transverse sinuses* and the *occipital sinus.* The two transverse sinuses occupy the attached margin of the *tentorium cerebelli.* After receiving the superior sagittal sinus, each transverse sinus leaves the tentorium cerebelli and curves medially and downward as right and left *sigmoid sinuses* to form the major portion of the *internal jugular veins* at the jugular foramina. Thus, the

Fig. 7–7.—The venous sinuses of the brain and their connections. The *arrows* indicate the course of spread of infection from the abscess in the scalp to the general systemic circulation. The complete course is demonstrated on the right side only (redrawn from Gardner *et al.*[2]).

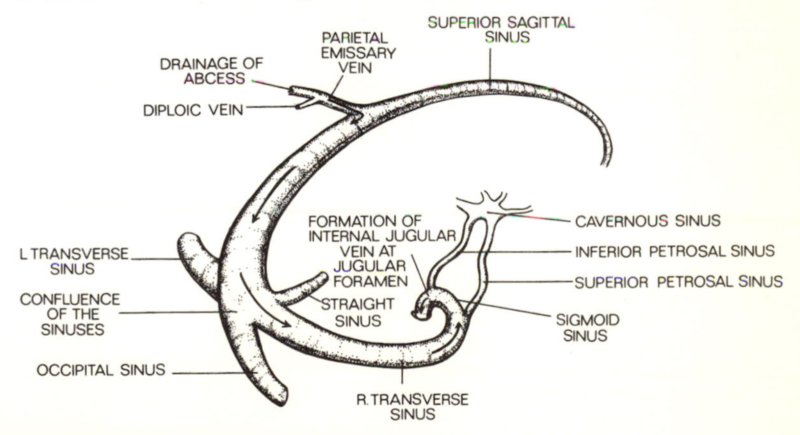

course of spread of infection in this case was from parietal emissary vein to superior sagittal sinus to right (and probably also left) transverse sinuses to sigmoid sinuses to internal jugular veins and from the latter to the general systemic circulation (see Fig. 7–7). The confluence of the sinuses communicates with the straight and occipital sinuses, whereas the transverse and sigmoid sinuses communicate with the *cavernous sinus* via the *superior* and *inferior petrosal sinuses*, respectively. It would be theoretically possible, although unlikely (due to the direction of blood flow), for the infection to reach the occipital or cavernous sinuses. Cavernous sinus infection, which is additionally serious due to its frequently manifested resistance to antibiotic treatment, was considered in Case 7–6.

Although blood normally flows away from the brain toward the scalp in the emissary veins, pressure on the abscess in this case forced the blood carrying the infection to flow in the reverse direction. The ease of this reversal resides in the fact that pressure in these veins and sinuses is low, blood flow is sluggish and valves usually are absent.

CASE 7–9: HEARING LOSS AND VESTIBULAR DISORDERS

History

A 62-year-old man went to his general practitioner with the primary complaint that he could not hear as well as normal in the right ear. This condition had developed slowly over the previous few months. A carefully taken history revealed that approximately 1 year prior to his realization that he was undergoing a hearing loss, the man had begun to observe "abnormal sounds" (buzzing, ringing) in the same ear. He was referred to specialists for a series of tests designed to determine the cause of the hearing loss. These included a series of modern methods of analysis of hearing function (audiology, audiometry), which showed that his hearing loss was general rather than of specific frequencies only. Examination by an otorhinolaryngologist revealed that the problem was not due to sclerosis of the middle ear ossicles or of the oval window (fenestra vestibuli), into which the footplate of the stapes fits and through which vibrations are transmitted to the inner ear. The man was fitted with a hearing aid and was scheduled for re-evaluation of the hearing problem in 6 months. Prior to that time, however, he began to experience the additional symptoms of vertigo, nausea and nystagmus, and he fell down several times while simply standing or walking. In addition, he observed a developing weakness in the muscles of the right side of the face and scalp, a reduction in taste sensation from the anterior two thirds of the tongue and reduced salivation and lacrimation on the right side.

When he finally returned to his physician for treatment, he was suffering from severe headaches and vomiting, and examination revealed choked optic discs. An x-ray examination demonstrated that the opening of the right internal auditory meatus was enlarged and the brain stem displaced to the left.

Questions

What single condition or phenomenon would account for all of the symptoms listed in this case?

Pinpoint the exact location of the problem and review the anatomy of this area, especially with regard to its relationships to the ear and to the brain stem.

Attempt to relate the individual or related groups of symptoms observed (e.g., vertigo, headaches) to the specific positions and relationships of the various structures in this area.

Would you expect this condition to be medically or surgically treatable?

Discussion

The symptoms in this case were the result of intracranial acoustic tumor or neurinoma of *cranial nerve VIII (acoustic, auditory, statoacoustic* or *vestibulocochlear nerve).* This type of tumor, which is thought to originate from a Schwann cell or cells, is characterized by a very slow rate of growth. In this case the tumor was located at the opening of the internal auditory meatus, which explains the enlargement of the opening observed on radiographs, i.e., by the time the tumor was diagnosed, it had eroded some of the bone in the area. Had the tumor been located within the bony canal itself (see below for discussion), the patient's complaints probably would have progressed more rapidly due to a more direct pressure on the nerves. In addition to cranial nerve VIII, *cranial nerve VII (facial)* along with its component, the *nervus intermedius,* also was compressed.

Both the facial and vestibulocochlear nerves arise from the groove between the pons and the medulla at the cerebellopontine angle. Both nerves travel laterally and enter the *internal auditory meatus.* The latter is a conspicuous opening in the posterior surface of the petrous portion of the temporal bone; this surface forms the anterior boundary of the posterior cranial fossa. The petrous portion of the temporal bone forms the boundary between the middle and posterior cranial fossae, houses the components of the internal ear and forms part of the boundaries of the middle ear. The vestibulocochlear nerve, as its name implies, is distributed to the vestibular or equilibratory portion and to the cochlear or auditory portion of the internal ear. These branches often are described functionally as separate entities even though they are components of one nerve. The *vestibular portion* sends impulses from the ampullary crests of the semicircular canals and from the maculae of the utricle and saccule to the brain stem. Compression of this portion of the nerve was responsible for the symptoms related to a dysfunction of equilibration, i.e., vertigo, nausea, nystagmus and falling. Analysis of the complexity of these connections, e.g., those related to eye movements, should be reserved for neuroanatomic discussions. The *cochlear portion* of the nerve carries impulses from the *spiral organ (of Corti)* to the brain. The initial symptom of the tumor (the "abnormal sounds," collectively known as *tinnitus,* that were heard by the patient) was the result of pressure on this division of cranial nerve VIII. Continuous pressure on the nerve would result in hearing loss (as was observed by the patient) and in eventual deafness.

The most important diagnostic evidence of the position of the tumor was the radiologic observation of the widening of the internal auditory meatus mentioned earlier. The physician also was able to determine, however, that the disorder was due to interruption of nerve impulses rather than to damage to either the middle ear or the cochlea by the fact that the facial nerve also was involved. The reduction in salivation, lacrimation and taste from the anterior two thirds of the tongue on the right side implicated the nervus intermedius (which gives rise to the greater petrosal nerve and the chorda tympani).

This strongly suggested, therefore, that a single tumor existed somewhere between the brain stem and the geniculate ganglion (at which point the previously named branches depart from the facial nerve), rather than an additional tumor of the seventh nerve at some point distal to the ganglion. The greater petrosal nerve exits from the temporal bone in a groove on the posterior slope of the middle cranial fossa, whereas the chorda tympani exits from the petrotympanic fissure in the infratemporal fossa. For a discussion of damage to the peripheral portion only of the facial nerve (i.e., distal to the stylomastoid foramen), see Case 5 of this chapter.

The slow growth of the tumor in the patient gradually displaced the brain stem to the left. It will be noted that ordinarily there is very little distance between the internal auditory meatus and the lateral surface of the pons (Fig. 7–8) and that the tumor, in addition to causing a compression of the nerves involved, also indirectly put them on the stretch. The initial displacement of the brain stem was asymptomatic, but continued pressure resulted in an impedance of the normal exchange of cerebrospinal fluid from the subarachnoid space. The resulting increase in intracranial pressure was responsi-

Fig. 7–8.—Relationships of the brain stem and certain cranial nerves to the cranial cavity. Note specifically the common exit of the *facial nerve* (cranial nerve VII) and *statoacoustic nerve* (VIII) at the *internal auditory meatus.*

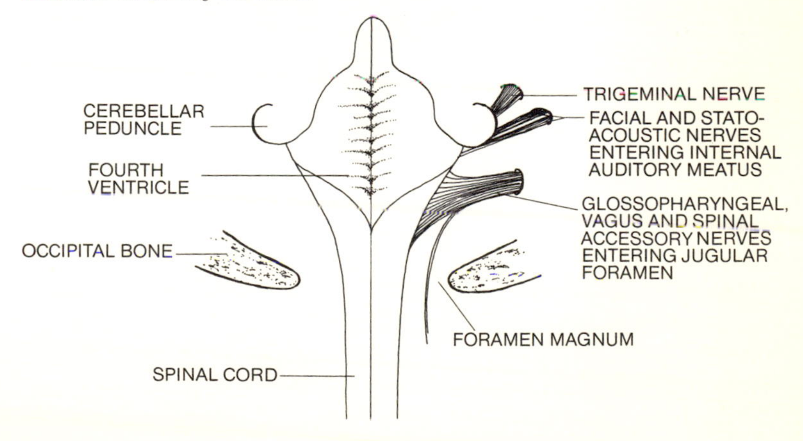

ble for the symptoms of headache, vomiting and choked optic discs. This particular tumor is of special clinical importance because it is often operable. In this case, particularly because the tumor occupied a position outside of the internal auditory canal, its extirpation was successful, and in time the patient's symptoms subsided to near normal limits.

CASE 7–10: NASOPHARYNGEAL TONSILLITIS

History

A 13-year-old boy was admitted to a hospital with severe tonsillitis involving primarily the nasopharyngeal tonsil. Preliminary examination revealed a temperature of 104 F, rapid pulse and respiration and considerable amounts of blood and pus in the sputum. The boy complained of a severe sore throat, bilateral earache and chest pains. A history revealed that the boy had had chronic symptoms of "adenoids" for the past 18 months and recurring bouts of tonsillitis for the previous 4 or 5 years. The former symptoms were, however, much milder than in the present episode. A throat examination revealed a purulent infection in the midline tonsillar area, with spreading to the posterolateral regions of the nasopharynx. The torus tubarius, which guards the opening to the auditory tube, was inflamed on both sides. Thoracic x-rays revealed a small accumulation of pus in the superior and posterior mediastina, posterior to the esophagus. The boy was immediately placed on large doses of wide-spectrum antibiotics given intramuscularly. His condition improved rapidly, and he was released from the hospital within 1 week of the initial crisis. He was back in the hospital 2 weeks later, however, for surgical removal of the tonsils, an action that precluded any further trouble of this nature.

Questions

How do you account for the patient's complaint of pain in both ears?

Review the anatomy of the nasopharynx, especially with regard to the important clinical relationships of the nasopharyngeal tonsil.

How did the infection spread from the nasopharynx into the superior and posterior mediastina?

What tissues or structures make up the pharyngeal wall?

Review the arrangement of the various layers of the deep cervical fascia that facilitated this spread.

Do fascial arrangements sometimes inhibit the spread of infectious processes also?

What prevented spread into the lateral portions of the neck in this case?

The student is encouraged to review the anatomy and relationships of the palatine tonsils at this time. What is included in the so-called "lymphatic ring" of the throat?

Discussion

The infection in the *nasopharyngeal (pharyngeal) tonsil* was severe enough to spread to both middle ear cavities via the auditory tubes (causing bilateral *otitis media*) and to actually invade and break through the posterior wall of the nasopharynx. The nasopharynx lies above the level of the soft palate, posterior to the nasal cavity with which it communicates through the choanae. Inferiorly the nasopharynx communicates with the oropharynx (which lies behind the oral cavity) via the *nasopharyngeal isthmus*. The single nasopharyngeal tonsil lies in the midline of the posterior wall of the nasopharynx and is bounded on either side by depressions known as the pharyngeal recesses. Anterior to the recess on each side, on the lateral wall, is the ostium of the *auditory (eustachian) tube,* which connects the cavity of the nasopharynx with that of the middle ear. The opening is bounded by a rounded elevation, the *torus tubarius* (or tubal elevation). A fold of tissue, the *salpingopharyngeal fold,* which covers the salpingopharyngeus muscle, descends on the lateral wall of the pharynx from the torus tubarius. This small muscle is active in swallowing, and it is in this process that tension is also put on the opening of the auditory tube, resulting in a momentary patency and an equalization of pressure between the pharynx (and, therefore, the exterior of the body) and the middle ear cavity. This explains the "popping" of the ears upon swallowing or yawning when altitude changes create pressure differentials in the middle ear cavity. If the ostium of the auditory tube is closed by swelling (such as in a cold or infection), pressure changes in the middle ear will result in a bowing in or out of the tympanic membrane. This can be the cause of considerable pain and even a temporary hearing loss.

An individual is spoken of as having the plural "adenoids" when the single nasopharyngeal tonsil is enlarged. The hypertrophy may become advanced to the point where it impinges on the soft palate, thereby reducing or cutting off air intake from the nose. If this condition becomes chronic, the individual is often spoken of as possessing *adenoid facies*, which is defined by Dorland[3] as "the stupid expression, with open mouth, seen in children with adenoid growths." In this particular case, the initial enlargement of the nasopharyngeal tonsil might have contributed to the severity of the boy's throat infection.

The walls of the pharynx may be subdivided into four distinct layers, which are, from the lumen outward, a *mucosa*, a *fibrous coat* (the *pharyngobasilar fascia* or *pharyngeal aponeurosis*), a *muscular coat* (with an outer circular and an inner, incomplete, longitudinal layer) and an external *fascial coat* (the *buccopharyngeal fascia*). The mucosa of the nasopharynx presents a pseudostratified, columnar, ciliated epithelium, the same as that of the respiratory portion of the nasal cavity. This undergoes transition to stratified squamous as the oropharynx is approached. The fibrous coat is attached superiorly to the base of the skull; it is anchored firmly in the posterior midline to the *pharyngeal tubercle* on the basilar portion of the occipital bone. The muscular coat of the nasopharynx consists largely of the *superior constrictor* muscle (circular coat), although the *stylopharyngeus* and *palatopharyngeus* muscles (longitudinal coat) contribute somewhat to its lateral wall. The superior constrictor muscle also attaches posteriorly to the pharyngeal tubercle but then loops *below* the auditory tube, leaving a gap that is devoid of muscle tissue. This gap, which corresponds to the site of the pharyngeal recess, lies on either side of the nasopharyngeal tonsil and is closed by the pharyngobasilar fascia. The buccopharyngeal fascia covers the muscles of the pharynx externally and blends superiorly with the fibrous coat.

The deep fascia of the neck *(deep cervical fascia)* is divided for descriptive purposes into three categories, the outer *investing fascia*, the *pretracheal (visceral) fascia* and the *prevertebral fascia*. Briefly, the investing layer surrounds the sternocleidomastoid and trapezius muscles on each side and then bridges the gaps between these muscles to form the roofs of the anterior and posterior triangles of the neck. The pretracheal fascia forms a sleeve around the trachea and esophagus in the neck and extends superiorly to form the external layer of the pharynx where it is called the buccopharyngeal fascia (see above). It also extends inferiorly into the thorax where it forms a fascial coat for the esophagus. The prevertebral fascia extends from the base of the skull to the tip of the coccyx. The cervical portion of this fascia lies anterior to the prevertebral muscles of the neck (longus colli or cervicis, longus capitis and rectus capitis anterior and lateralis), the scalenus muscles, and splits to enclose the posterior neck muscles. In many places in the neck these fascial layers are in apposition with each other. They are not bound together, but rather are free to glide against each other, which greatly enhances the freedom of movement of the head and neck. This situation, however, creates potential *fascial clefts* or spaces between the layers, and it is through these clefts that infections may spread from one part of the body to another.

In this particular case the infection in the nasopharyngeal tonsil cut its way through the pharyngobasilar fascia in the area of the pharyngeal recess, which is deficient in muscle fibers. After traversing the buccopharyngeal fascia (part of the visceral or pretracheal fascia) and an additional named layer of the deep cervical fascia, the *alar fascia*, the infection occupied the *retropharyngeal fascial cleft* (space) between the abovementioned layers and the prevertebral fascia (Fig. 7–9). The infection was then able to spread easily into the thorax, since this space continues inferiorly first into the superior and then into the posterior mediastinum between the esophagus and vertebral bodies. The infectious spread was checked laterally in the neck due to the fusion of the pretracheal, alar and prevertebral fasciae with that forming the carotid sheath. Thus, in certain instances the fascial arrangements of the body facilitate this sort of spread, whereas in others they tend to inhibit or even to fence it off altogether. The spread of infections via fascial clefts, once a very serious surgical problem, is now, in the age of antibiotics, a relatively rare phenomenon.

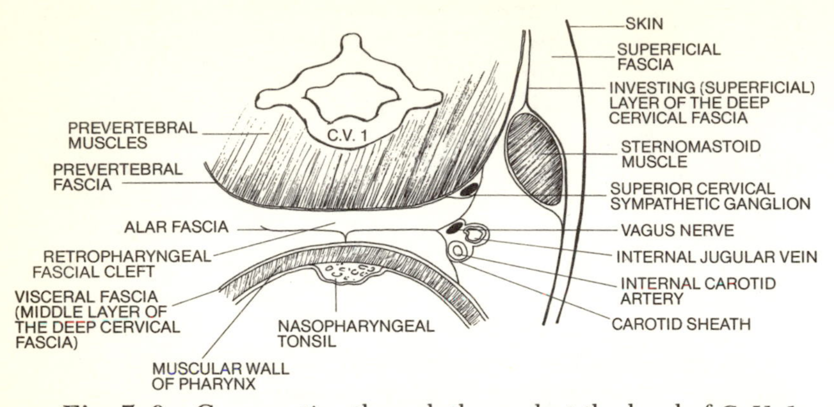

Fig. 7–9.—Cross-section through the neck at the level of *C. V. 1*. Observe that an infection emanating from the *nasopharyngeal tonsil* could rupture posteriorly into the *retropharyngeal fascial cleft* and from there spread inferiorly into the posterior mediastinum. Note also the relationship of the *internal carotid artery* to the superior *cervical sympathetic ganglion* (the "potential" fascial clefts have been exaggerated intentionally in this illustration).

The *lymphatic (Waldeyer's) ring* consists of the nasopharyngeal, palatine and lingual tonsils, which surround the *faucial isthmus*, leading from the oral cavity into the pharynx. It has been suggested that this structure assists in guarding against the ingestion of harmful material and acts as a partial barrier to the spread of infections. On the other hand, tonsils may become the seat of chronic infections. Whether or not to routinely remove children's tonsils has been, and undoubtedly will continue to be, a controversial subject.

CASE 7–11: RHINITIS AND SINUSITIS

History

A 27-year-old woman developed a chronic nasal infection *(rhinitis)* as the result of a severe head cold. The infection, which was manifested as sneezing and an alternately stuffed or runny nose, was annoying but not debilitating. With time, however, the woman began to have severe headaches in the area of the forehead bilaterally above the eyebrows. In addition, she felt increased stuffiness and pressure deep to the middle face area accompanied by occasional discharges of pus from the nose. She tried several over-the-counter preparations that relieved her discomfort temporarily. Her symptoms, however, recurred whenever the medicine wore off, and eventually she began to feel pain lateral to the nose on each side accompanied by toothaches in various upper teeth. Since her condition continued to deteriorate and was beginning to affect her general well-being, both physically and emotionally, she consulted her physician, who diagnosed her condition as chronic paranasal *sinusitis*. She was treated as an outpatient and released and, as a result, her condition gradually improved until she was symptom free.

Questions

Review the basic gross anatomy, histology and developmental anatomy (both pre- and postnatal) of the various paranasal sinuses.

Why do you think the patient's pain occurred originally in the forehead and later spread to areas bilateral to her nose?

By what means and to what areas do the sinuses normally drain?

What normal functions would you ascribe to the paranasal sinuses?

How do you explain the patient's toothaches?

What do you think would be the most effective means of treating this condition?

Discussion

An understanding of the lateral wall of the nasal cavity is essential in this discussion since the paranasal sinuses are outgrowths of this cavity. The sinuses communicate with the cavity through ostia and in certain cases through ducts. The respiratory mucosa of the nasal cavity (sometimes referred to as a mucoendosteum) is continuous with the mucosa of the paranasal sinuses through the ostia and ducts. The mucosae are lined by a pseudostratified, ciliated, columnar epithelium that contains numerous glands. The lateral walls of the cavity are characterized by the presence of *superior, middle* and *inferior conchae* or turbinate bones that project medially from its surface into the cavity. These bones are thought to create turbulence in the inspired air, thus warming cold air through increased contact with the richly vascular mucoendosteum.

The paranasal sinuses open into the *sphenoethmoidal recess,* which lies above the superior concha, and into the *superior* and *middle meatus,* which lie below the superior and middle conchae, respectively. Specifically, the *sphenoidal sinus* drains into the sphenoethmoidal recess; the *posterior ethmoidal sinuses* into the superior meatus; and the *frontal, maxillary* and *anterior ethmoidal sinuses* into the middle meatus (Fig. 7–10). The nasolacrimal duct, which is not part of this system but which drains tears from the lacrimal apparatus of the eye, empties into the inferior meatus.

Development of the paranasal sinuses begins during the fourth month of fetal life, but most of their development occurs after birth. The proportion of the nasal cavity and maxillary sinuses to the cranial portion of the head is considerably less in the newborn infant than in the adult. The frontal sinuses, which are often small or absent at birth, invade the frontal bone postnatally, and the sphenoidal sinus invades the sphenoid bone during childhood.

The bilateral *maxillary sinuses* are the largest of the paranasal sinuses and are situated within the body of the maxilla bones. The floor of the sinus is formed by the alveolar process of the maxilla. This close relation of sinus to upper teeth helps to explain the toothache often experienced in maxillary sinusi-

Fig. 7–10.—The lateral wall of the nasal cavity demonstrating the openings of air cells and sinuses into the various meatus. The *arrows* show the course of spread of infection from the *frontal sinus* to the *maxillary sinus.*

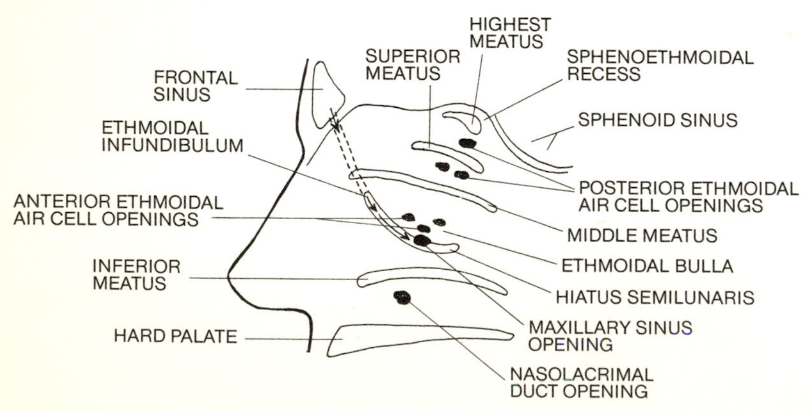

tis as well as secondary maxillary sinus infections frequently observed in patients in whom the primary focus was in an upper tooth. The *ethmoidal sinuses* (or *air cells*), which number anywhere from four to around 15, are small cavities associated with the frontal, maxilla, lacrimal, palatine and sphenoid bones. The bilateral *frontal sinuses* are asymmetric cavities within the frontal bones. They commonly extend backward into the roof of the orbit and, therefore, pain caused by an infection in this sinus is often felt above the eye, usually at the medial aspect of the eyebrow. The *sphenoidal sinus* is usually single, but it may be partitioned by a shelf of bone. It is highly variable in both size and shape and is closely related to various components of the brain.

Special attention needs to be given to the pattern of drainage of the various sinuses into the middle meatus of the nasal cavity (see Fig. 7–10). This meatus presents an elevation known as the *ethmoidal bulla*, which receives the openings of the anterior ethmoidal sinuses. Inferior to the bulla is a curved groove known as the *hiatus semilunaris*, into which the maxillary sinus drains (at about the junction of the anterior two-thirds with the posterior one-third). The hiatus leads anterosuperiorly to the *ethmoidal infundibulum*, which receives the openings of the frontal sinus and a few of the anterior ethmoidal sinuses. This arrangement explains the observation in this particular case that the infection had spread from the frontal sinus, where it originated, to the maxillary sinus, which served as a reservoir for a secondary infectious process. Drainage of pus from the frontal sinus occurred by gravitational flow down the infundibulum into the hiatus and from there into the maxillary sinus through its ostium. In addition to gravity, the paranasal sinuses are drained by the ciliary action of their epithelia, by a negative pressure created during inspiration and by a suction caused by blowing the nose. Congestion of the mucosa caused by an infection may result in occlusion of one or more paranasal sinus ostia. The infection in such a circumstance would continue to develop within the now-closed space, resulting in pressure and usually headache, as was described in this case. An additional complication is the possible loss of cilia, which could result in permanent impairment of the normal flow of mucus from the sinus.

The normal functions of the paranasal sinuses are a matter of dispute among investigators in this area. It is quite likely that sufferers of chronic or recurrent sinusitis feel the problems associated with these structures outweigh any possible benefits derived from them. Teleologic arguments have been advanced to support one or another theory as to function; for example, some investigators feel their function is to pneumatize or lighten the facial bones, although the amount of weight saved is probably insignificant in light of the over-all weight of the skull. Others feel that the sinuses add *resonance* to the voice, since twangy, dull nasal sounds are characteristic of patients in whom the ostia are occluded and/or in whom the sinuses may be filled with purulent material or mucus.

Foreign material in the sinuses may be detected by transillumination, e.g., by placing a strong light within the mouth and observing the maxillary sinus from in front, or by radiologic technics with or without injection of radiopaque dyes. All of the paranasal sinuses are susceptible to the spread of infections from the nose. The maxillary sinus presents a specific problem with regard to sequestering infectious materials because the normal position of its ostium is frequently too far superior for effective drainage, i.e., a shelf or baffle of bone exists that facilitates a retention of fluid. In such cases, the surgeon may have to artificially create a new opening by puncturing either the lateral wall of the inferior meatus or the canine fossa on the anterior surface of the maxilla. Irrigation of the sinuses also is performed when normal drainage is inadequate. Medical treatment consists of prescription of decongestants and vasoconstrictors to attempt to recreate normal drainage from the sinuses and prescription of antibiotics in the presence of purulent infections.

CASE 7–12: KNIFE WOUND IN THE NECK

History

A young man was taken to the emergency room of the local county hospital with a knife wound in the right side of the neck suffered in a gang fight. The knife blade pierced the skin and sternomastoid muscle in a horizontal plane at a point approximately 4 cm directly posterior to the angle of the mandible. Some bleeding was evident from the wound, but it was not severe enough to warrant surgical intervention. Compresses were applied to the area of the incision and, once the bleeding was under control, the skin was stitched and the patient released. Upon his return 1 week later to have the stitches removed, the following symptoms were observed: (1) the patient's tongue deviated markedly to the right upon protrusion; (2) the floor of the mouth on the right side was relaxed and lengthened: (3) the hyoid bone was tilted slightly from the horizontal plane, the right side being a little higher than the left and (4) a generalized weakness of the infrahyoid muscles on the right side was present. There was no sensory loss from the tongue or from the skin of the neck, and taste sensation from the tongue was judged normal. In addition, function of the sternomastoid and platysma muscles was unimpaired. On subsequent visits the patient's symptoms were less and less pronounced until, at the end of a year, function was largely restored in the tongue and associated structures.

Questions

Did the knife enter one of the specifically named triangles of the neck? If so, which one?

Damage to what structure would account for the symptoms listed above?

Why did the tongue deviate toward the injured side?

How do you account for the symptoms related to the hyoid bone and infrahyoid muscles? Why were the latter not totally denervated?

How do you account for the fact that only minimal bleeding occurred from the site of the wound?

Review the relationships of cranial nerves IX, X, XI and XII to the arteries and other structures of the neck.

Why do you think the patient recovered use of the affected structures in this particular case?

Discussion

The lateral surface of the neck is quadrangular and roughly square (Fig. 7–11, A). The superior border is formed by the lower border of the body of the mandible and by a line continued posteriorly to the occipital region of the skull. The inferior border is formed by the clavicle (middle third), the posterior by the anterior edge of the trapezius muscle and the anterior

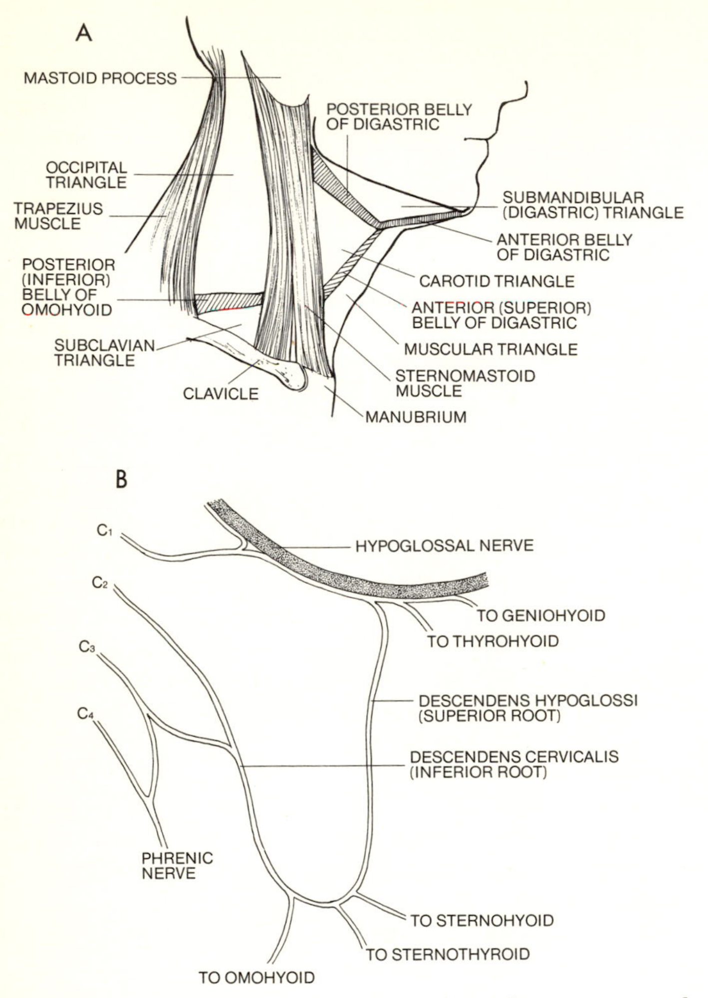

Fig. 7–11. – **A,** the triangles of the neck. **B,** the ansa cervicalis of the right side of the neck.

by the anterior midline of the neck. The sternomastoid muscle bissects this quadrangular area from the posterosuperior angle to the anteroinferior angle and thus divides the neck into *anterior* and *posterior triangles*. Each of these triangles is further subdivided into additional smaller triangles, named for descriptive purposes according to their contents. Thus the posterior triangle contains an *occipital* and *subclavian triangle,* whereas the anterior triangle contains *submandibular, carotid* and *muscular triangles.* Since the knife in this patient directly pierced the sternomastoid muscle, it did not enter one of the specifically named triangles of the neck.

The structure that was injured and thus resulted in the observed symptoms was the *right hypoglossal nerve* (cranial nerve XII). This nerve supplies both the intrinsic and extrinsic muscles of the tongue with motor fibers. The *intrinsic muscles,* which originate and insert in the substance of the tongue itself, are named descriptively as *superior* and *inferior horizontal, vertical* and *transverse* bundles. The *extrinsic muscles,* which arise from outside of the tongue but insert on it, are the *palatoglossus, styloglossus, hyoglossus* and *genioglossus.* In general, the name of the muscle indicates its origin, insertion and action. For example, the styloglossus muscle originates on the styloid process ("stylo-") and inserts on the tongue ("-glossus"). It should be apparent from the positions of these structures that contraction of the muscle would result in *retraction* of the tongue. The genioglossus arises from the superior genial tubercle near the symphysis menti of the mandible and therefore protrudes the tongue. It is important for the student to note that paralysis or total relaxation of the genioglossus (as in general anesthesia) might allow the tongue to move posteriorly and thus block the oropharynx, resulting in possible suffocation. All of the muscles of the tongue occur bilaterally, and all are supplied by the hypoglossal nerve with the exception of the palatoglossus, which is supplied by the pharyngeal plexus of nerves (largely cranial nerve XI). The protruded tongue deviates to the injured side because the denervated muscles are unable to "push" or protrude the tongue on an equal basis with the unaffected muscles. The result is that the tip of the tongue is swung around, in this case to the right.

The symptoms related to the positioning of the hyoid bone resulted from the fact that the hypoglossal nerve carries fibers from C_1 and sometimes C_2 (Fig. 7–11, B), which supply the *geniohyoid muscle* (one of the suprahyoid muscles) and the *thyrohyoid muscle* (an infrahyoid muscle). These fibers simply travel with the hypoglossal nerve but are not strictly speaking a part of it; they would, however, obviously be cut along with the hypoglossal nerve. The geniohyoid pulls the hyoid bone forward and, as a result, shortens the floor of the mouth, whereas the thyrohyoid depresses the hyoid bone. The symptoms noted were therefore the result of reduced opposition by the denervated muscles to the other muscles that act upon the hyoid bone. In addition, the hypoglossal nerve gives off a *superior root* (also known as the *descendens hypoglossi*) of the *ansa cervicalis*. The fibers in this nerve are similarly derived from C_1, often with a contribution from C_2. The ansa cervicalis supplies the remaining infrahyoid muscles (omohyoid, sternothyroid, sternohyoid). The fact that the ansa (ansa = "loop") is completed by an *inferior root* (also known as the *descendens cervicalis*), derived usually from C_2 and C_3, explains why the infrahyoid muscles displayed weakness, but not total paralysis, in this patient.

The aforementioned nerves are motor in composition. There was no sensory loss to the neck area because the nerves supplying this modality are derived from the cervical plexus (anterior primary rami of $C_1 - C_4$), which is independent of the ansa cervicalis. There was no loss of general sensation from the tongue, since this is carried in the lingual branch of the mandibular division (V_3) of cranial nerve V (anterior two-thirds) and in the glossopharyngeal nerve (posterior one-third). Similarly, there was no loss of taste sensation, which is carried in the chorda tympani branch of the facial (anterior two-thirds) and in the glossopharyngeal (posterior one-third). The platysma muscle, which often covers the area in question, was not affected since it is supplied by the cervical branch of the facial nerve; the sternomastoid muscle similarly remained intact since its nerve supply, the accessory cranial nerve (XI) was unharmed.

Fortunately for the patient, the knife wound was not of sufficient depth to enter either the internal or external carotid arteries, which lie just medial (deep) to the hypoglossal nerve. Indeed, the blade did not totally sever the nerve; this probably explains why the patient recovered almost completely from the initially observed muscle denervation, i.e., the approximation of the nearly severed nerve ends facilitated regeneration. The relationships of cranial nerves IX–XII to the internal and external carotid arteries and other structures of the neck are of considerable importance. The glossopharyngeal nerve (IX) and the pharyngeal branch of the vagus (X) are lateral to the internal carotid but medial to the external carotid artery. The superior laryngeal branch of the vagus lies medial to both arteries. The main trunk of the vagus descends posteriorly to the internal and then to the common carotid artery. The accessory nerve (XI) crosses the posterior triangle of the neck diagonally from the jugular foramen superiorly to pass under cover of the anterior border of the trapezius muscle. The hypoglossal nerve (XII) is the most superficial of the cranial nerves in this area. As it descends from the hypoglossal canal, it hooks around the occipital artery and lies external to both the internal and external carotid arteries.

CASE 7–13: FACIAL PAIN

History

An elderly woman presented with severe, sometimes excruciating paroxysms (recurrences at various intervals) of pain, which she felt beneath the left eye (radiating to the lower eyelid and lateral side of the nose and upper lip), in the left superior alveolar area and over the zygomatic and temporal areas on the left side. She described the pain as "stabbing" or "cutting" and noted that it disappeared altogether between paroxysms. She felt that the pain was "set off" somehow by touching certain areas of the face or especially by the act of chewing on the left side of the mouth. Symptoms of autonomic nervous system irritation often accompanied the pain. For example, she demonstrated lacrimation, injection of the conjunctival blood vessels and flushing of the affected part of the face (other patients with this affliction often additionally demonstrate salivation during painful attacks).

A complete neurologic examination was performed and, aside from the pain and related symptoms, no other disturbances were evident. For example, there was no altered sensibility in either the affected area or any other portion of the face, and no neuritis was evident. A history revealed that she had undergone dental extractions from the left maxillary area of the mouth several months prior to onset of the pain but, aside from this possible relationship, the etiology of the condition was not readily apparent. Because of the severity of the pain (the woman had contemplated suicide during various paroxysms), she was referred immediately to a neurosurgeon, who was able by various procedures to eventually control the condition.

Questions

What specific structure is directly involved in this patient's condition?

Review in as much detail as you can the anatomy of this structure and its ramifications.

Discuss the connections of this structure and others related to it with autonomic nervous system fibers that might explain why various other symptoms accompanied the pain.

From your knowledge of anatomy and physiology, what procedures might be used by the neurosurgeon to alleviate the patient's pain? Which might be temporary and which permanent?

Discussion

The symptoms listed in this patient are characteristic of the condition known as *trigeminal neuralgia* or *tic douloureux*. It is manifested as severe pain in one or more divisions (ophthalmic, maxillary and/or mandibular) of the *trigeminal nerve* (cranial nerve V). In this particular instance the pain was felt in the left *maxillary division* (V_2) of the nerve. The etiology of this condition, which often causes great suffering and distress, is uncertain. It is possible that the earlier extraction of the patient's teeth contributed to the onset of the painful paroxysms, but such speculations have not been proved statistically. The condition is most prevalent in middle-aged and elderly persons. Controversy exists as to whether the origin of the condition lies in central pathways or in the peripheral nerve fibers themselves. Some investigators believe it is due to age-related changing anatomic relationships between the internal carotid artery and the trigeminal (semilunar) ganglion.

The trigeminal nerve is one of the most complex of the cranial nerves with regard to its distribution (see Case 7–4 for an additional discussion of this nerve). It arises from the side of the pons as two roots, a larger sensory one and a smaller motor one. The joined roots pass from the posterior cranial fossa to the middle by crossing the petrous portion of the temporal bone under cover of the tentorium cerebelli. The nerve is expanded into the *trigeminal ganglion* on the anterior slope of the petrous temporal (posterior wall of the middle cranial fossa) lateral to the dorsum sellae of the sella turcica just above the foramen lacerum. The ganglion contains the cell bodies of the general somatic afferent fibers, which reach it via its three named nerve trunks or divisions (ophthalmic, maxillary and mandibular). The mandibular nerve also carries the motor fibers (special visceral efferents) to the muscles of mastication.

The left maxillary nerve (Fig. 7 – 12), the affected division in this case, enters the middle cranial fossa from the pterygopalatine fossa through the foramen rotundum. This nerve gives origin to the following peripheral branches, which explain the distribution of pain in the patient's face: (1) *Posterior superior alveolar branches,* which descend in canals on the posterior aspect of the maxilla and supply the posterior maxillary teeth, the gums in that area and the maxillary sinus. (2) The *zygomatic nerve,* which passes through the inferior orbital fissure and then runs a short distance on the lateral wall of the

Fig. 7–12.—The course and distribution of the maxillary division (V_2) of the left *trigeminal nerve* (cranial nerve V).

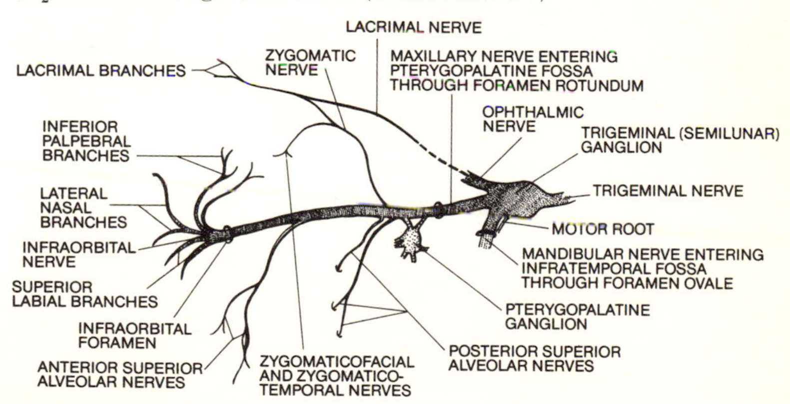

orbit before dividing into the *zygomaticotemporal* and *zygomaticofacial branches*. The former emerges from the orbit through a bony foramen to supply the skin of a portion of the temple, whereas the latter emerges over the malar eminence (prominence of the cheek) to supply the skin in that area. (3) The *infraorbital nerve*, usually considered the continuation of V_2; this nerve traverses the infraorbital canal, gives off *middle* and *anterior superior alveolar branches*, which supply the respectively named teeth and gums, and then emerges from the infraorbital foramen on the anterior face. From this point, branches ascend to the lower eyelid *(inferior palpebral branches)*, extend medially to the nose *(lateral nasal branches)* and descend to the upper lip *(superior labial branches)*. Anyone who has had an infraorbital nerve block for dental work on the upper teeth is familiar with the distribution of the branches of this nerve. In addition to these branches, V_2 also gives off a meningeal branch and roots, which communicate with the pterygopalatine ganglion.

The autonomic symptoms observed (e.g., lacrimation, flushing of the face) can be explained partially by the relationship of both sympathetic and parasympathetic fibers with the sensory fibers of the left V_2. The student should recall that the pterygopalatine ganglion is suspended from V_2 as the latter passes through the pterygopalatine fossa (see Fig. 7–12). The cell bodies of preganglionic sympathetic neurons lie in the intermediolateral column of the lateral horn of the upper thoracic spinal cord. They ascend in the cervical sympathetic chain and synapse on cell bodies of postganglionic neurons in the superior cervical ganglion. The latter form a plexus on the internal carotid artery. These fibers leave the artery as it approaches the foramen lacerum and enter the pterygoid canal as the *deep petrosal nerve*. The cell bodies of preganglionic parasympathetic fibers lie in the superior salivatory nucleus of the brain stem and run in the *greater petrosal nerve* (a branch of the facial nerve). These fibers join the deep petrosal nerve in the pterygoid canal, the combined nerves now being referred to as the *nerve of the pterygoid canal* (vidian nerve). This nerve proceeds anteriorly through the canal to join the pterygopalatine ganglion. The sympathetic fibers, being postganglionic, simply pass through the ganglion, whereas the preganglionic parasympathetic fibers synapse on postganglionic neurons in the ganglion. From this point, both types of autonomic fibers are distributed with the branches of V_2. A connection between the zygomatic nerve and the lacrimal nerve in the orbit allows distribution of the autonomic fibers to the lacrimal gland. The sympathetic fibers are largely vasomotor, whereas the parasympathetics are secretomotor to the various glands of the area. Salivation would most frequently be associated with neuralgia of the mandibular nerve (V_3), whereas involvement of the ophthalmic nerve would be the major contributor to injection of the conjunctival blood vessels.

Initial treatment in this patient involved the injection of alcohol (phenol also has been used in recent years) into the left V_2 at the foramen rotundum. When more than one nerve trunk of the trigeminal is afflicted, the substance often is injected through the foramen ovale into the trigeminal ganglion itself. This procedure was immediately effective in alleviating the patient's facial pain, but her symptoms began to return several months after the injections, becoming severe within a year. Therefore, the surgeon performed a "retrogasserian root section" (cutting the trigeminal nerve between the pons and the trigeminal ganglion). This resulted in permanent anesthesia over the entire side of the woman's face. This procedure is conditionally undesirable in that loss of sensation in the cornea presents the hazard of ensuing keratitis unless special precautions are taken. Tractotomy procedures (medullary or trigeminal) avoid this complication. Other therapeutic measures in this condition involve "decompression" of the ganglion by incising the overlying dura in the middle cranial fossa, or "compression" (crushing or clamping) of the nerve root or one of its divisions. Occasionally, partial deep sensibility returns to the face after nerve root section. Some investigators have attempted to explain this phenomenon by suggesting that fibers from the cervical plexus, facial nerve or sympathetic chain may grow into the denervated area.

CASE 7–14: INTERRUPTION OF BLOOD SUPPLY TO THE BRAIN

History

A 77-year-old woman complained to her physician of occasional dizziness of some duration and of a recent fainting spell, which had prompted her visit. A routine physical examination demonstrated normal cardiac size and function, blood counts within normal ranges and no respiratory, ophthalmic or equilibratory disorders. There was some evidence of generalized atherosclerotic disease, however. Her physician, therefore, decided to run a series of cerebral angiographic examinations (roentgenographic examination of the cerebral blood vessels following injection of radiopaque materials into the common carotid artery of one or both sides). The tests, which also included an analysis of the branches of the external carotid arteries, revealed thrombi in the left internal carotid artery at the level of the carotid canal and in the right facial artery at the point where it crosses the inferior border of the mandible. Both vessels were totally occluded. Since the patient was not seriously debilitated and showed no evidence of ischemia over the right side of the face, the physician concluded that the clots had developed over a long period of time. This gradual formation had allowed adequate collateral circulation to develop through anastomoses between branches of the external and internal carotid arteries (internal carotid clot) and between branches of the facial artery and branches of the internal carotid and maxillary arteries (facial artery clot).

Questions

Describe the normal blood supply to the brain and face and discuss as many collateral pathways as you can that would carry arterial blood to the areas distal to each thrombus.

How do you explain the late onset of the patient's fainting spell?

Discussion

Basically the arterial supply to the brain is derived from the two *internal carotid arteries* and the two *vertebral arteries.* The internal carotids, one of the two terminal branches of the common carotid arteries, ascend in the neck without giving any branches and enter the bony carotid canal at the base of the skull. The course of the artery through the floor of the skull en route to the cranial cavity is complex, and the student is encouraged to consult a standard anatomy text for a description. Within the cranial cavity each internal carotid artery gives off an *anterior cerebral artery*, a *middle cerebral artery* and a *posterior communicating artery* (Fig. 7–13, A). The two

Fig. 7–13.—A, the arterial circle of the brain (circle of Willis). **B,** the *facial artery*.

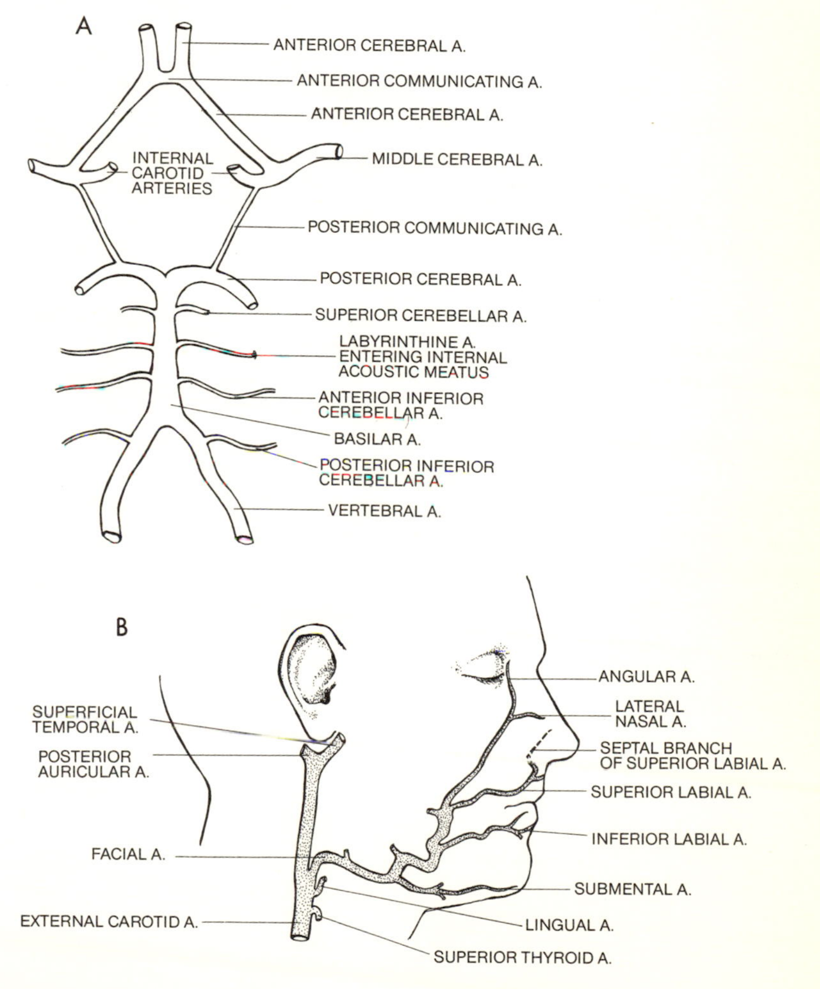

anterior cerebral arteries are connected across the midline by an *anterior communicating artery*. The vertebral arteries, branches of the first part of the subclavian arteries, ascend through the foramina transversaria of C. V. 1 through C. V. 6. After a tortuous course through the suboccipital region, each artery enters the cranial cavity through the foramen magnum. At about the inferior border of the pons, the two vertebral arteries unite to form the *basilar artery*. The latter ascends on the inferior aspect of the pons, supplying branches to this structure, the cerebellum and the internal ear. After a short course, the basilar artery terminates by dividing into the two *posterior cerebral arteries*. Each posterior cerebral artery is connected via the posterior communicating artery with the internal carotid artery of the respective side. Thus, a continuous anastomotic arterial circle is formed around the base of the brain. This structure is known as the *circulus arteriosus* or *circle of Willis*, and it is the major source of collateral circulation in cases of blockage of one or more of the major vessels that contribute to its formation. For example, blood could be supplied to the area distal to the thrombus in the left internal carotid artery via the following courses: (1) right internal carotid → right anterior cerebral → anterior communicating → left anterior cerebral → left internal carotid (to site of blockage), or (2) right and left vertebrals → basilar → left posterior communicating → left internal carotid. It is important to note that under such conditions, blood would flow retrograde to the normal direction in certain vessels, e.g., in a portion of the left anterior cerebral artery cited in course 1 above.

There are a number of smaller and less prominent anastomoses that occur between the internal and external carotid arteries that are nevertheless very important in establishing collateral circulation around the clot. Numerous branches of the *ophthalmic artery* (a branch of the internal carotid siphon) anastomose with branches of the *external carotid artery*. Examples of these anastomoses are (1) the supraorbital branch of the ophthalmic with the anterior branch of the superficial temporal, (2) the posterior ethmoidal branch of the ophthalmic with the sphenopalatine branch of the maxillary, (3) the dorsal nasal branch of the ophthalmic with the angular and lateral nasal branches of the facial and (4) the recurrent branch of the lacrimal (also a branch of the ophthalmic artery) with the middle meningeal branch of the maxillary artery. Some or all of these anastomoses may be present in any given situation. They probably exist under normal conditions as capillary or arteriolar anastomoses, but they have the potential of greatly increasing in size upon demand.

The *facial artery* usually is the third branch to arise from the external carotid in the neck, although it may spring from a common trunk with the lingual artery. It runs upward and forward across the carotid triangle deep to the stylohyoid and digastric (posterior belly) muscles, grooves the posterior surface of the submandibular gland and then winds around the inferior border of the mandible at the anterior margin of the masseter muscle. It then ascends in a tortuous course across the face toward the medial angle of the eye, where it is known as the *angular artery*. The facial artery also possesses numerous anastomoses that provide collateral circulation in the event of a thrombosis (Fig. 7–13, B). Most of these occur with other branches of the external carotid artery but, as noted above, the angular and lateral nasal branches anastomose with the dorsal nasal branch of the ophthalmic artery, and thus there is an ultimate communication with the internal carotid artery. In this case, blood would flow in the opposite direction to that of the internal carotid artery discussed above.

It is apparent that, if both thrombi in this patient had occurred on the same side, this particular collateral pathway would have been closed at both ends, although the circle of Willis would feed into it between the clots. Other examples of anastomoses in this situation are (1) the superior labial branch of one facial artery with the superior labial branch of the other facial; (2) the superior labial branch of the affected facial with the sphenopalatine branch of the right maxillary (an interesting sidelight here is that this anastomosis is the most frequent site of nosebleeds; thus, pressure on the superior labial artery often reduces or arrests blood flow from the nose); (3) the angular and lateral nasal branches of the facial with the infraorbital branch of the maxillary artery and (4) the inferior labial and submental branches of the facial artery with the mental artery, which is a branch of the inferior alveolar artery, itself a branch of the maxillary artery.

The late onset of the patient's fainting spell was the result of the very gradual formation of the thrombus and the gradual progression of atherosclerotic disease. Whereas a younger person might be able to tolerate complete occlusion of one or more of the blood vessels of the head, the patient's advanced age precluded such resistance. Hamby[4] cites the example of a 48-year-old woman who survived bilateral ligation of both internal carotid arteries without apparent impairment of vision.

CASE 7–15: EYE PAIN AND REDUCTION OF VISION

History

An 83-year-old man called his physician in the evening complaining of severe pain in the left eye accompanied by blurred vision in that eye. He reported that the pain was of such sudden onset and severity that he collapsed to the floor. Vomiting ensued shortly thereafter. He described the pain as being located mainly in the eye but with spreading also to the eyebrow, cheek and temple of the left side. He was instructed to proceed immediately to the hospital, where upon his arrival he was examined by an ophthalmologist. The eye was severely congested, the cornea demonstrated a hazy opacity and the pupil was slightly dilated and inactive. The anterior chamber appeared reduced in the anteroposterior dimension. The eyeball was tender to the touch and felt rigid and hard upon palpation. A history revealed that the attack was probably precipitated by overexertion (the patient had been moving furniture). In addition, he stated that he had experienced visual disturbances in the form of blurred vision and colored haloes around lights at night for several years prior to this episode. The condition was treated both medically and surgically, which relieved the symptoms and prevented their recurrence, respectively.

Questions

Identify this disorder and discuss generally its etiology and classifications.

Describe briefly the gross anatomy of the entire eye.

Describe also the histology and histophysiology of the specific portions of the eye that are directly related to this condition.

What specific medical act is *always* to be avoided in suspected cases of this disorder, and why?

What is the medical treatment of choice and what is the rationale for this treatment?

What is the specific goal of follow-up surgery?

Discussion

The patient in this case suffered an attack of *acute congestive glaucoma (narrow angle glaucoma).* The etiology of this disease rests in mechanisms (see below) that result in increased intraocular pressure from the norm of approximately 20 mm Hg to as high as 90 mm Hg. With this increase in pressure, the retinal artery, which enters the eyeball at the optic disc, is compressed, thus reducing the normal supply of nutrients to the retina. The result may be permanent atrophy of the retina and/or optic nerve, leading to irrevocable blindness. Consequently, this condition is one of the true emergencies in ophthalmology. Other forms of glaucoma are classified as *infantile, simple glaucoma (open angle glaucoma)* and *secondary glaucoma* (the result of some other conditions of the eye, e.g., trauma, intraocular tumors or retinal vein thrombosis).

The outer layer of the eye, consisting of the white *sclera* and the transparent *cornea,* is a tough, fibrous protective layer. The middle layer or *choroid* is highly vascular and pigmented to form a dark chamber. The innermost layer is the light-sensitive *retina* (see Case 8–2). The retina is continuous with the *optic nerve,* which exits from the posterior aspect of the eye en route to the brain. The *lens* (Fig. 7–14) is suspended from the *ciliary body* just posterior to the *pupil,* the latter a circular opening in the *iris.* The optical system of the eye, which forms an image on the retina, consists of the cornea, the lens and the fluid media or humors contained within the globe. The *aqueous humor,* as its name reveals, is watery; it fills the cavities anterior to the lens. These are the *anterior chamber* in front of the iris and the *posterior chamber* behind it. The space posterior to the lens and lying internal to the retina is occupied by the *vitreous humor,* a gel that resembles raw egg white.

The junction between the cornea and the sclera is known as

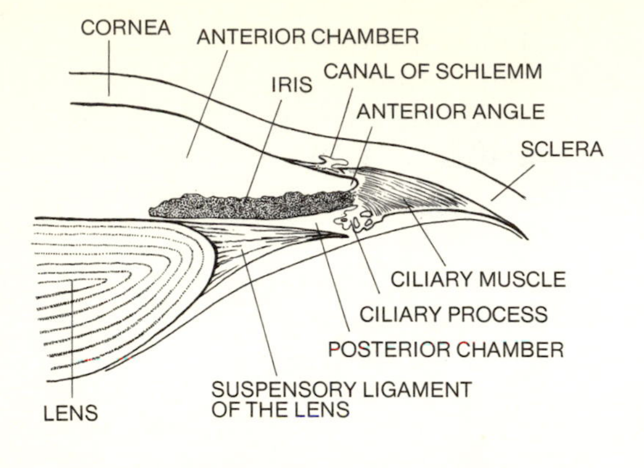

Fig. 7–14.—The relationships at the *anterior angle* of the eye.

the *limbus*. Internal to this point is the *anterior angle* of the eye formed between the iris and the cornea (see Fig. 7–14). In this angle is found a trabecular meshwork that consists of a series of parallel, flattened lamellae. The lamellae consist basically of a collagenous core covered by endothelium. The channels between the lamellae are known as the *spaces of Fontana*. The *canal of Schlemm* encircles the eye between the sclera and the anterior angle. It divides into branches that soon recombine only to divide again. It communicates centrally with the spaces of Fontana and peripherally by means of 20–30 small branches with the *anterior ciliary veins* of the adjacent scleral tissue.

The aqueous humor is secreted at a rate of about 0.012 cc per minute by the epithelium of the ciliary body through both passive and active transport mechanisms. This process is similar to the formation of cerebrospinal fluid by the choroid plexuses in the ventricles of the brain. After it is secreted by the ciliary process, the aqueous humor flows from the posterior chamber (where it forms an interface with the vitreous humor), between the ligaments of the lens and through the pupil into the anterior chamber. It is picked up at the anterior angle by the spaces of Fontana, from there drains into the canal of Schlemm and finally leaves the eye by the anterior ciliary veins. Other outflow channels have been postulated, e.g., choroidal blood vessels. The aqueous humor supplies the lens and cornea with nutrients and helps to maintain the shape of the eye.

A very efficient, but poorly understood, mechanism maintains a nearly constant intraocular pressure throughout life. The following regulatory mechanisms have been postulated: (1) increased intraocular pressure, causing distention of the spaces of Fontana, thus enhancing drainage of aqueous humor from the eye and restoring normal pressure; (2) variations in the rate of formation of aqueous humor by the ciliary process maintaining a near-constant pressure; (3) changes in resistance in the canal of Schlemm depending on changes in intraocular pressure or (4) blood pressure variation in the anterior ciliary veins, thus increasing or decreasing resistance in response to changes in intraocular pressure. Glaucoma occurs when, for whatever reason, the rate of outflow of aqueous humor drops below the rate of its formation. Congestive glaucoma suggests an acute impedance of drainage of aqueous humor as a result of a narrowing of the anterior angle of the eye. A dilated pupil, as observed in this patient, is suggestive of this process, as is the reduced anteroposterior dimension of the eye. Further dilation of the pupil with atropine (which is used in standard ophthalmologic examinations) always is contraindicated in suspected cases of glaucoma, since this would tend only to compound the problem and lead to further damage to the eye.

Treatment is therefore aimed at constricting the pupil in an attempt to widen the anterior angle, thereby reducing congestion and enhancing the outflow of aqueous humor from the eye. Miotic drugs, such as esserine and pilocarpine, are used for this purpose. Surgical treatment, in the form of follow-up iridectomy with or without the creation of a new drainage channel, usually prevents recurrence of acute congestive glaucoma.[5]

REFERENCES

1. Martin, H.: Radical neck dissection, Ciba Clinical Symposia 13: 103, 1961.

2. Gardner, E., *et al.*: *Anatomy* (3d ed.; Philadelphia: W. B. Saunders Co., 1969).
3. Dorland: *Illustrated Medical Dictionary* (24th ed.; Philadelphia/London: W. B. Saunders Co., 1965).
4. Hamby, W. B.: *Intracranial Aneurysms* (Springfield, Ill.: Charles C Thomas, 1952).
5. Jackson, C., R. S.: *The Eye in General Practice* (5th ed.; Edinburgh/London: E. & S. Livingstone, Ltd., 1969).

Chapter 8

NERVOUS SYSTEM

CASE 8–1: NEUROLOGIC DISORDER OF THE FACE

History

A routine chest x-ray of a 62-year-old man demonstrated a small solid tumor on the left side of the superior mediastinum. Specifically the tumor occupied a position at the left side of the body of T. V. 1, just anterior and inferior to the articulation of the first rib. During surgery for removal of the tumor, the team of physicians observed that it had invaded the local area to a greater extent than previously expected. Therefore, considerable dissecting and cleaning of adjacent structures was performed during surgery, and any associated mediastinal lymph nodes were removed. The latter were biopsied and found to be negative, i.e., no local metastases were evident. The surgery was therefore apparently successful, and the patient was released from hospital several days later. Within a short period of time, however, the man began to complain of the following physical symptoms, all related to the left side of the face and scalp and over the majority of the surface of the left upper extremity: (1) anhidrosis; (2) reduction of sebaceous secretion; (3) flushing or redness of the skin (due to vasodilation of the surface blood vessels) and (4) absence of function of the arrectores pilorum muscles (small smooth muscle bundles that attach to the hair roots and produce the characteristic "goose pimples" or "goose flesh" upon contraction). In addition, he demonstrated ptosis of the left upper eyelid and miosis of the left pupil. The physician also observed that the pulse rate was slightly lower than normal.

Questions

What specific structure was damaged during surgical removal of the patient's tumor? At what level was it injured?

What are the three basic divisions of the nervous system?

Review the anatomy and basic physiology of the two subdivisions of the autonomic nervous system.

Can central control be exerted over this system?

Why was the man's head involved if the lesion was in the upper thoracic area?

Would a lesion lower down in this area create any different or additional symptoms?

Why is heart pain perceived in (referred to) the left upper extremity as well as locally (to the substernal area)?

Would you expect to observe any other symptoms as a result of this injury?

Discussion

In this case, the left *sympathetic chain* (trunk) was incised during surgery at a level just inferior to the stellate ganglion. The *stellate ganglion,* also known as the cervicothoracic ganglion, is formed by the union of the inferior cervical and the first thoracic sympathetic ganglion. No damage to the peripheral nerves themselves was detectable.

The mammalian nervous system is subdivided into the *central nervous system,* which includes the brain and spinal cord; the *peripheral nervous system,* which contains the cranial and spinal nerves; and the *autonomic nervous system,* which is further divided into the *sympathetic* and *parasympathetic systems.* Functionally, the sympathetic component tends to bring about bodily functions that prepare an individual for an emergency or for quick action (the so-called "fight or flight" reaction). Thus, the heart beats faster, the pupils dilate, blood pressure is elevated, peripheral blood vessels constrict, sweating is increased, hair stands up, bronchi dilate and actions of the stomach and intestines are decreased or inhibited. The parasympathetic system, which is largely antagonistic to the sympathetic, is involved with conservation of bodily resources and with restoration to homeostatic conditions. Therefore, as a result of parasympathetic stimulation, the heart rate slows, the pupils constrict, blood pressure drops, the stomach and intestines increase their activity and glandular secretion is enhanced. The term "autonomic" implies a lack of conscious control, although higher centers definitely exert control over this system, as evidenced in emotional disturbances. The functional components of the autonomic nervous system are the *general visceral efferent* (motor) fibers, whereas general visceral afferent (sensory) fibers, which travel in many of the same pathways, are not, strictly speaking, considered a part of this system.The sympathetic division also is referred to as the *thoracolumbar outflow,* since the cell bodies of its preganglionic neurons lie in the intermediolateral column of the lateral horn from thoracic cord segment one to approximately lumbar cord segment two (T_1-L_2). The parasympathetic division is also known as the *craniosacral outflow* due to its origin in cranial nerves III, VII, IX and X and in sacral nerves S_2, S_3 and S_4. In each division there is a *preganglionic* and a *postganglionic* fiber; one of the basic differences between the two is that the sympathetic pre- and postganglionic fibers may be approximately the same length, whereas in the parasympathetic division the preganglionic neurons usually are very long and the postganglionics short, for the most part lying in the wall of the innervated organ itself. There also are differences in the chemical constitution of the synaptic transmitters of the two divisions, a subject for physiologic discussion.

The preganglionic sympathetic fibers pass from the lateral horn of the spinal cord into the ventral root and thence into a spinal nerve (Fig. 8–1). The fibers then pass from the nerve into a sympathetic ganglion, there being one such ganglion on each side for each spinal nerve between T_1 and L_2 segments of the cord. Each ganglion is connected to the nerve by two roots, the *white* and *gray rami communicantes.* The white

Fig. 8–1.—Thoracic spinal cord segment showing the pattern and distribution *(arrows)* of *pre-* and *postganglionic neurons* of the sympathetic division of the autonomic nervous sytem. Preganglionic fibers also may enter a splanchnic nerve, and postganglionic fibers also may ascend or descend in the *sympathetic trunk.*

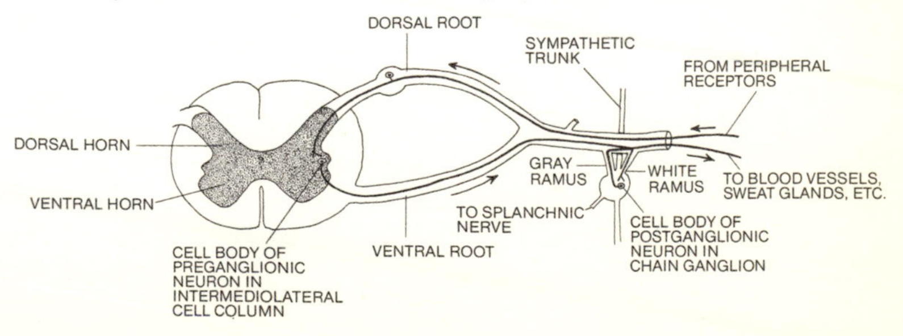

ramus communicans is so named because it carries the preganglionic sympathetic fibers, which are myelinated to the ganglion (the myelin gives a white appearance to the ramus in fresh specimens). The preganglionic fiber synapses on the cell body of the postganglionic neuron in the chain ganglion. The gray ramus carries the nonmyelinated postganglionic fiber from the ganglion. The student should be aware that the white ramus actually is distal to the gray one and, as a result, the fibers cross each other on their way to and from the ganglion (see Fig. 8–1). Sympathetic ganglia also occur above and below T_1-L_2 (i.e., cervical, lumbar, etc.), and all are connected by a chain (trunk). Only those in the direct outflow area, however, possess white rami as well as gray; those above and below possess only gray rami communicantes (postganglionic fibers). In addition, the latter ganglia are less regular in both form and position than those in the outflow area (T_1-L_2).

Once the preganglionic sympathetic fiber has reached the ganglion, it can take one of the following courses: (1) It can synapse in the ganglion at its same level, the postganglionic fiber then passing into the respective spinal nerve (to supply, for example, the smooth muscle of blood vessels, the sebaceous and sweat glands and the arrectores pilorum muscles of the thoracic and abdominal walls); (2) it can pass up or down the chain (to synapse, for example, in the superior cervical ganglion in the neck, the postganglionic fibers of which supply most of the thus innervated structures of the head); (3) it can enter one of the *thoracic splanchnic nerves* (greater splanchnic T_5-T_9, lesser splanchnic T_{10}, T_{11} and lowest splanchnic T_{12}), which pass through the diaphragm and into the abdominal cavity to synapse on postganglionic cell bodies located in one of the collateral ganglia in that area (i.e., celiac, aorticorenal, superior and inferior mesenteric, the fibers of which supply abdominal organs) or (4) it may combine any of the above (for example, synapse in a chain ganglion and then traverse the greater splanchnic nerve to supply the lower portion of the esophagus). It should be noted, however, that by far the majority of fibers in the thoracic splanchnic nerves are preganglionic. The student should be aware that the viscera of the thoracic and abdominal cavities also receive sympathetic innervation from direct postganglionic branches of chain ganglia in the respective area. For example, the heart receives postganglionics from thoracic cord levels T_1-T_4. The lumbar splanchnic nerves in the abdomen supply a portion of the large intestine, and small nerves from the cervical ganglia supply the heart.

The exception to the rule of long preganglionic fibers and short postganglionics in the parasympathetic division lies in the presence of four sets of named ganglia in the head that are the sites of synapse of these fibers. These are the *ciliary ganglion* (to structures of the eye), *pterygopalatine ganglion* (to glands of the nasal cavity and palate and the lacrimal gland), *submandibular ganglion* (to the submandibular, sublingual and lingual glands) and *otic ganglion* (to the parotid gland). It should be recalled that parasympathetic innervation to glands is secretomotor in nature, e.g., causing lacrimation or salivation. The vagus nerve (vagus means "wandering") supplies parasympathetic fibers to the thorax and most of the abdomen. The remainder of the abdominal viscera is supplied with parasympathetic fibers from sacral nerves S_2, S_3 and S_4 by way of the pelvic splanchnic nerves or nervi erigentes.

The man's head was involved, creating the symptoms known as the *Horner syndrome* (see Case 7–7), because the entire sympathetic outflow to the left side of the head was cut off when the left sympathetic chain was incised (there is no preganglionic sympathetic outflow superior to T_1). The preganglionic sympathetic neurons to the head are derived from the upper thoracic level (approximately T_1-T_4). They synapse in the superior cervical ganglion, and postganglionic fibers are then distributed to the head along the course of various blood vessels. A lesion lower down, e.g., at T_6 or T_7, would not result in the symptoms of the Horner syndrome, since it would be below the level of the preganglionic outflow to the head. Instead, we would expect to observe problems associated with abdominal structures that ordinarily would be supplied by fibers running in the greater splanchnic nerve.

The fact that the heart receives direct postganglionics from T_1-T_4, as well as those from the middle cervical and stellate ganglia, explains why denervation in this patient was not complete on the injured side. In other words, only those fibers superior to the lesion were rendered devoid of impulses. Sympathetic innervation to the upper limb is derived primarily from T_1 (and often from T_2), postganglionics finding their way into all of the major nerves of the brachial plexus (spinal nerves C_5 and C_6 receive fibers from the middle cervical ganglion, whereas the remainder receive theirs usually from the stellate ganglion). Thus, as in the head, complete denervation of the upper limb also occurred simply by transecting the sympathetic chain below the stellate ganglion. If any sympathetic innervation had remained in the extremity, it would have to have been derived from the second thoracic ganglion.

Frequently, when afferent fibers from different sources reach the cord at the same level, pain in a viscus such as the heart may be perceived as also coming from a surface structure supplied by the same spinal nerve. This phenomenon is known as *referred pain*, the details and hypotheses of which should be deferred to a course in neuroanatomy. Nevertheless, general visceral afferent fibers from the heart reach the spinal cord at the same level as general somatic afferent fibers from the upper limb (T_1–T_2). The conjunction of these fibers somehow translates pain from the heart to the shoulder and upper limb as in the classically recognized occurrence in acute myocardial infarction. Sympathetic innervation to the lungs and the upper esophagus was probably not interrupted in this case; even if injury to these nerves had occurred, it probably would not have been clinically detectable in these organs.

CASE 8–2: LOSS OF VISION FOLLOWING HEAD TRAUMA

History

An elderly woman, while cleaning her kitchen, bumped her head sharply on the undersurface of a cupboard. The blow produced considerable local pain and a lump but did not break the skin. It gave her a headache, however, so she took two aspirin tablets and rested for several hours. The headache abated and no symptomatology, other than a tender scalp, persisted. However, 6 days later she observed "sparks" or "flashing lights" in the right eye, accompanied by showers of dark objects that floated from superior to inferior across the visual field. She later described this to her ophthalmologist as resembling a "curtain coming down over my eye." This onset of symptoms was accompanied by a gradual but painless loss of vision in that eye. Examination through the dilated pupil revealed loss of the normal red reflex from the retina and a "billowed" appearance of a portion of this structure accompanied by a greater than normal degree of convolution of the retinal blood vessels. The latter also were noticeably darker than normal. She was treated surgically, and rest was prescribed for the affected eye. Due to the rapidity with which she sought treatment, she recovered nearly complete vision in the eye.

Questions

What is your diagnosis in this case?

How is this condition related to the blow on the woman's head and to her age?

What other situations might predispose to it?

Describe the retina from an embryologic, gross anatomic, histologic and neurologic standpoint.

What types of treatment are used in the correction of this disorder?

Discussion

The woman suffered a *simple retinal detachment* that resulted from the jolt to her head. It is common for the major symptoms to show up some days after the initial trauma. This is probably explained by the fact that the initial blow caused a slight tearing of the retina, which allowed fluid to accumulate between its two layers (see below). If enough fluid accumulates, the inner layer detaches from the outer one, resulting in the billowing effect observed by the ophthalmologist and the symptoms experienced by the patient. The incidence of retinal detachment increases with advancing age, a phenomenon probably associated with age-related pathologic changes in the retina specifically and the eye in general. Retinal detachment also occurs more frequently in myopes, presumably because the eye is larger than normal and the retina may be degenerative.[1] In addition, Jackson notes, "Retinal detachment is one of the possible delayed complications of intraocular surgery, especially cataract surgery." These complications may result from a change in the fluid dynamics of the eye due to the removal of the lens. Nevertheless, retinal detachment may occur in young normal individuals as a result of head trauma (e.g., in boxers or football players). A *secondary* or *solid detachment of the retina* is one in which the elevation of the retina from its bed results from a malignant neoplasm, e.g., a malignant melanoma of the choroid or a metastatic tumor from perhaps the prostate or breast, or from some inflammatory process.

This case has been included in the chapter on the nervous system rather than in the previous chapter because the retina really is an extension of the brain itself, the so-called optic "nerve" in truth being a brain tract rather than a typical cranial nerve (see Case 7–15 for a more detailed discussion of the anatomy of the eye). The eye develops as a double-walled cup attached to the brain by a stalk (Fig. 8–2). The outer layer of the cup becomes the *pigmented layer* (commonly called the pigmented epithelium) of the retina, whereas the inner layer develops into the *nervous layer* of the retina (often considered the retina proper). It is these two layers that are separated in retinal detachment.

The pigmented layer consists of a single layer of short, fat cells, the outer portion of which contains the nucleus. The

Fig. 8–2.—The optic cup and developing *lens*.

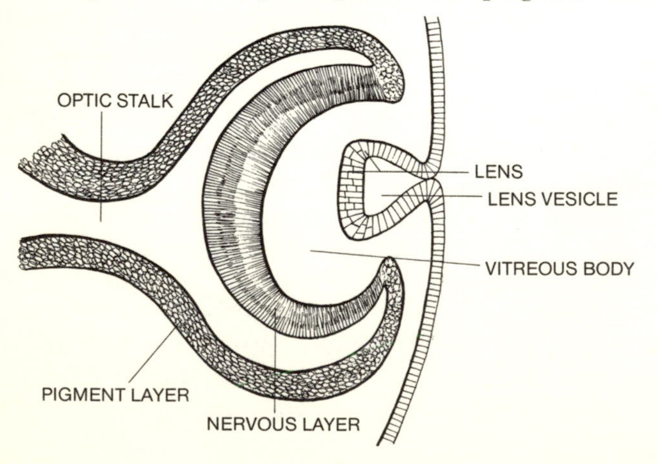

inner layer is filled with pigmented granules, which extend down into cytoplasmic processes that envelop the ends of the rods and cones in the nervous layer. The latter layer is much more complicated than the pigmented one, consisting of as many as nine or more subdivisions, depending on the descriptive source. For our purposes, we need to concentrate basically on three of these sublayers. The first (from outside in) consists of the light receptor cells, the *rods* and *cones*. Rods are sensitive to low levels of light intensity, whereas cones are color sensitive. Both of these cell types synapse on cells of the second sublayer of the nervous retina consisting of the *bipolar cells*. These cells in turn synapse on the cells of the third sublayer, the *ganglion cells*, the axons of which comprise the optic nerve. It should be noted that light, upon striking the retina, has to pass through the layers of ganglion and bipolar cells before hitting the actual photoreceptors, the rods and cones.

The retina is covered externally by the *choroid*, which in turn is covered by the *sclera*. Internally, it is in contact with the *vitreous body*, which assists in maintaining the shape of the eye. The retina terminates anteriorly at the *ora serrata*. At the site of exit of the optic nerve, there is a depression known as the *optic disc*. Since it contains no rods or cones, it frequently is referred to as the "blind spot." Lateral to the optic disc is the *macula lutea*, a yellowish spot that presents a small depression, the *fovea centralis*. The macula possesses a high density of cones but no rods; it is the area of the retina with the greatest visual sensitivity. The retina is supplied by the *central artery of the retina*, a branch of the *ophthalmic artery*. It is one of the few examples in the body of a true end artery, and sudden occlusion results in immediate and irrevocable blindness.

Operation is the treatment of choice in simple retinal detachment. Until recently diathermy was used to produce an inflammatory reaction and subsequent adhesion of the retina to the choroid. This has been replaced by the more sophisticated technics of photocoagulation (by laser beam), cryosurgery (application of an intensely cold rod to the sclera just external to the retinal tear) or plastic procedures (e.g., scleral resection).

CASE 8–3: LOSS OF CUTANEOUS SENSATION OVER THE SHOULDERS

History

An 18-year-old male had noticed for several weeks a gradual reduction in cutaneous sensation over the shoulders and upper limbs. Specifically, he had gradually lost pain and temperature sensibility from these areas while, curiously, maintaining the ability to detect touch and deep sensibility as well as the ability to discriminate shapes and textures. As the condition progressed, he noticed that he frequently burned himself with cigarettes without feeling heat or pain. Eventually he began to detect motor disability in the upper limbs, manifested as paresis leading to paralysis and characterized by muscular atrophy and weakness, loss of tone and weakened or abolished deep tendon reflexes in the extremities. These clinical signs began on the ulnar sides of the hands and spread to the radial sides, to the forearms and finally to the thorax. Physical examination revealed the presence of congenital anomalies in the form of supernumerary nipples and facial asymmetry. Radiologic examination demonstrated an abnormal widening of the central canal in the cervical portion of the spinal cord and an increase in the normal transverse diameter of the cervical cord shadow. The patient was given radiation therapy, which slowed the progress of the disease considerably, but the sensory and motor losses were permanent.

Questions

What is the name of this condition?

Explain from an anatomic standpoint why only pain and temperature sensations were lost and not the other sensory modalities.

Would you expect the losses to be bilateral or unilateral, and why?

How do you account for the motor losses experienced by the patient as the condition progressed?

What might be some possible causes of this disease?

Discussion

This condition is known as *syringomyelia* and is caused by cavitation of the spinal cord, originating in the central canal and progressing peripherally. The cavities most frequently occur in the cervical enlargement of the cord, as occurred in this patient, and are found less often in the lumbar enlargement. The cavitation is accompanied by a glial reaction (gliosis), which spreads along the advancing margin of the syrinx.

In very general terms, a transverse section of the human spinal cord reveals a central area of *gray matter*, shaped something like a butterfly, and surrounding *white matter* (Fig. 8–3). In the center of the gray matter is the small *central canal*. The gray matter consists largely of neuronal cell bodies with some nerve fibers, whereas the white matter is made up mainly of fiber tracts. The gray matter consists of *dorsal* and *ventral horns* with an *intermediate zone* intervening between them. Above and below the central canal are the *dorsal* and *ventral gray commissures*, respectively. The white matter consists of the *dorsal, lateral* and *ventral columns* or *funiculi* on each side. Beneath the ventral gray commissure is an area where nerve fibers decussate (cross from one side to the other) known as the *ventral white commissure*. In the area between the dorsal horn and the surface of the cord is the *dorsolateral fasciculus* (or *zone of Lissauer*).

The unipolar cell bodies of the afferent neurons carrying pain and temperature sensation from the periphery are located in the *dorsal root ganglia*. The central processes of these cells

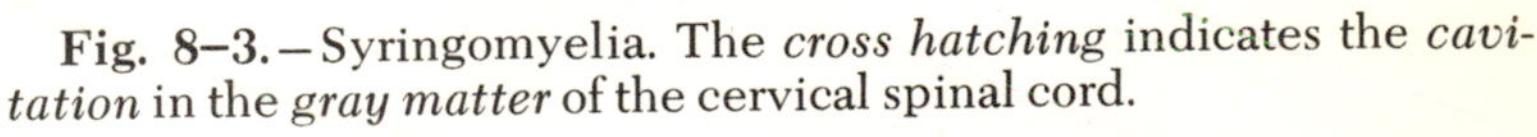

Fig. 8–3. – Syringomyelia. The *cross hatching* indicates the *cavitation* in the *gray matter* of the cervical spinal cord.

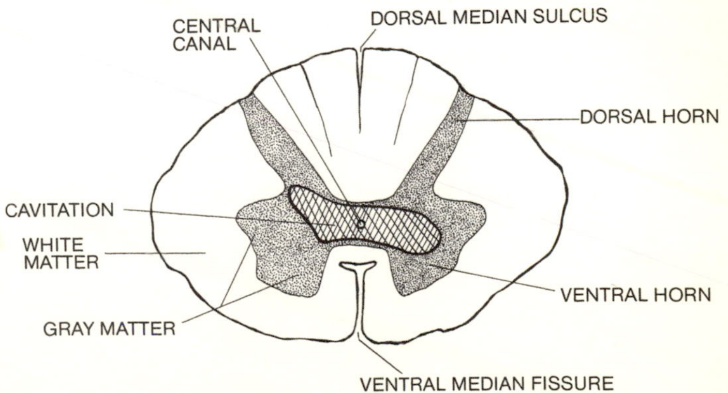

enter the dorsolateral fasciculus of the spinal cord via the lateral portions of the dorsal roots. The processes divide into ascending and descending fibers that traverse one or two cord segments and then synapse on cells in the *substantia gelatinosa,* located in the dorsal horn. The processes from these cells (or perhaps from intercalated neurons upon which these cells synapse) decussate in the ventral white commissure and come to lie bilaterally near the periphery in the ventrolateral regions of the lateral columns as the *lateral spinothalamic tracts.* These tracts, carrying pain and temperature sensations, ascend to the thalamus (see Case 8–4).

Cavitation of the spinal cord, beginning at the central canal, therefore affects the decussating pain and temperature fibers first (see Fig. 8–3). Since this occurred in the cervical cord (usually C_8–T_1), the patient first noticed loss of sensation in the areas subserved by these cord segments (shoulders, upper limbs, upper thorax). These are bilateral losses since the cavitation destroys fibers derived from both sides of the body. The pattern of sensory loss is referred to as a "yoke" or "jacket" type. The modalities of proprioception, tactile discrimination, three-dimensional sensation and light touch were not affected since the nerve fibers carrying these sensations do not cross the midline. As the cavitation progresses, the ventral horns become affected and, since they contain the cell bodies of motor neurons, the symptoms of muscle paralysis, etc. (as listed above) become apparent.

The etiology of syringomyelia is unknown, although it is more frequent in individuals who demonstrate some congenital anomaly or anomalies.[2] Barr[3] notes certain hypotheses based on neuropathologic studies: "It has been suggested, for example, that the pulsatile circulation through the choroid plexuses produces pressure waves in the cerebrospinal fluid, that these are carried into the central canal from the fourth ventricle, and that in rare instances the eventual result is cyst formation and surrounding gliosis in the region of the central canal. Another suggestion is that when the cerebrospinal fluid pressure in the spinal canal is increased for any reason, the pressure is transmitted into the central canal of the cord through perivascular extensions of the spinal subarachnoid space."

CASE 8–4: CORDOTOMY FOR INTRACTABLE PAIN

History

An elderly woman was diagnosed as suffering from cancer of the large intestine. A laparotomy revealed widespread metastases to abdominal and pelvic organs and, as a result, her condition was diagnosed as terminal. No further surgery was performed at this time. Radiation therapy and chemotherapy were prescribed primarily as palliative measures. Initially, she was relatively free of pain, but within a few weeks it became severe and finally intractable, responding poorly to large doses of morphine. As a result, laminectomy and cordotomy were performed, which immediately relieved the great majority of the severe pain. Some of the discomfort recurred before the patient's death, about 9 weeks after the latter operation.

Questions

What is a laminectomy?

How do pain fibers from the abdominal and pelvic viscera reach the spinal cord?

What major spinal cord tract is cut in a cordotomy?

Would this process need to be performed unilaterally or bilaterally in this case?

Discuss the formation of this tract in the spinal cord and trace its projection to the higher centers of the brain.

Why was all of the patient's pain not alleviated by cutting this tract?

Discussion

A *laminectomy* (see also Case 2–8) involves the surgical excision of the posterior arch of one or more vertebrae. It will be recalled that the vertebral arch is attached to the body of the vertebra by the *pedicles,* which constitute pillars on either side of the *vertebral foramen.* In a typical vertebra the *laminae* extend from the attachment of the *transverse processes* to the pedicles to the *spinous process* in the midline. The two pedicles, laminae and the base of the spinous process, therefore, form the vertebral foramen (combination of the vertebral foramina of the total vertebral complement is known as the *vertebral canal*) and house the spinal cord and cauda equina (see Case 2–8 for a more complete description of the spine). Laminectomy, therefore, exposes the spinal cord in the vertebral canal.

General visceral afferent fibers from the small and large intestine course with the sympathetic plexuses of the abdomen and enter the spinal cord through the dorsal roots of segments T_8–L_2. In general, it may be said that afferent fibers from abdominal viscera enter the cord anywhere from T_6–L_3. The pelvic organs are supplied by the *pelvic splanchnic nerves (nervi erigentes)*, which originate from sacral cord levels S_2–S_4. Pain fibers from the pelvis probably enter one or more of the following: (1) lumbar splanchnic nerves, (2) superior hypogastric plexus, or (3) pelvic splanchnic nerves. The cell bodies of the afferent neurons lie in the dorsal root ganglia. The tract that is cut in a *cordotomy* is the *lateral spinothalamic tract*, which carries pain and temperature sensations to the thalamus for relay to higher centers. Specifically, these fibers terminate in the *posterior ventrolateral nucleus of the thalamus*. From the thalamus, fibers radiate through the internal capsule for eventual projection to the postcentral gyrus of the cerebral cortex. This is the main sensory or somatesthetic area of the cerebral cortex and, although some pain may be perceived at the thalamic level, true pain consciousness occurs in the cortex. The pathway of fibers from the dorsal root to the spinothalamic tracts (e.g., zone of Lissauer to substantia gelatinosa and then decussation in the ventral white commissure) was covered in Case 8–3.

The lateral spinothalamic tract lies in the ventral portion of the lateral funiculus of the spinal cord (Fig. 8–4). A *somatotopic* arrangement of fibers exists, i.e., fibers from the inferior aspects of the body occupy the dorsolateral aspect of the tract, whereas the more superiorly originating fibers lie in a ventromedial position. Some investigators believe the temperature fibers are more dorsal than those subserving the sensation of pain. Cordotomy involves cutting the spinothalamic tract(s) in order to eliminate the pain sensation that originates below the level of incision. Since the fibers cross the cord in the ventral white commissure, a unilateral cordotomy on the right side would alleviate pain (and temperature) sensation from below

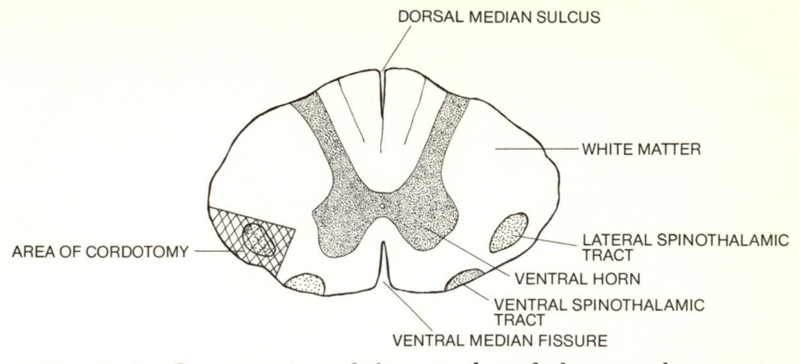

Fig. 8–4.—Cross-section of the spinal cord showing the area incised in a cordotomy in order to interrupt the pain fibers ascending in the *lateral spinothalamic tract.*

that point on the contralateral side of the body. Since afferent fibers from the abdominal and pelvic viscera enter the cord bilaterally for the most part, in this case a bilateral cordotomy would have been required. Performed superior to the sixth thoracic segment of the cord, it would have relieved pain from all areas of the body that provide afferent fibers to the cord from below this point (abdomen, pelvis, lower limbs).

The important landmarks in performing a cordotomy are the *dentate ligaments,* which consist of numerous toothlike projections of the pia mater. These arise in the midlateral plane of the spinal cord and pierce the arachnoid laterally to attach to the dura mater and, therefore, assist in anchoring the spinal cord within the vertebral canal. In a laminectomy the dorsal half of the cord is superficial to the dentate ligaments and the ventral half deep to them. Since the spinothalamic tracts lie in the ventrolateral portion of the cord, the incision must be made deep to the dentate ligaments. Many neurosurgeons believe that, since this is a terminal procedure, the greatest degree of success is obtained by cutting nearly the entire ventrolateral quadrant of the cord rather than simply nicking the surface to a depth of only 2 or 3 mm. In a bilateral cordotomy then, the entire ventral half of the cord is likely to be incised. Both the spinothalamic and trigeminothalamic tracts may be cut by a single incision in the lateral surface of the medulla or midbrain. If this is done, pain and temperature sensations are lost over the entire face, neck, trunk and limbs of the opposite half of the body.

The observation that not all of a patient's pain always is relieved in these procedures or that it tends to recur has been the subject of a great deal of discussion. The following tracts have been suggested, therefore, as possibly carrying pain fibers: (1) spinocervical, (2) spinospinal, (3) spinoreticular and/or (4) uncrossed spinothalamic. The fact that the spinothalamic tracts are largely, if not entirely, crossed, however, whereas others are not, is made clear in the *Brown-Séquard syndrome* in which the spinal cord undergoes hemisection. In this case, pain and thermal sensations are lost contralaterally below the lesion, whereas proprioception, two-point discrimination and vibratory sensations are lost *ipsilaterally* below the level of the lesion.

CASE 8–5: ENLARGED HEAD IN A CHILD

History

It was noted by the pediatrician of a 2-month-old male child that he did not seem to be gaining weight, was hyperirritable at times and extremely lethargic at others, seemed dehydrated and frequently emitted a high-pitched cry. The head seemed to be too large for the body, the latter appearing frail and somewhat wasted. A history revealed that the birth had been difficult and with a breech presentation. Physical examination demonstrated a slight craniofacial disproportion, the dome of the skull being relatively enlarged. As a result, the eyes and ears gave the appearance of being low set. In addition, the eyes were displaced downward so that the pupils were almost on a line with the margin of the inferior eyelid. The scalp was thin and shiny, the veins enlarged and the anterior fontanel increased in size. There was no evidence of optic nerve atrophy or papilledema. The child was treated surgically and, with the appropriate supportive management, recovered to some degree from the disorder. Ten years later, however, the problem was still apparent.

Questions

What is the name of this disorder, how is it classified and what are its possible causes?

From an anatomical standpoint discuss the origin, the course and the reabsorptive mechanism of the cerebrospinal fluid.

Describe very briefly the pathogenesis of this condition.

Discussion

The symptoms of this condition are, of course, characteristic of *hydrocephalus* ("water on the brain"), the diagnosis made by the pediatrician after careful testing. Hydrocephalus is classified as *communicating* in nature, i.e., the arachnoid villi and perivascular spaces are somehow obstructed, and *non-communicating,* i.e., the ventricular system is blocked by one means or another. These characteristics may be listed as two possible causes of the disorder, i.e., impaired venous absorption or obstruction of the cerebrospinal fluid pathways. A third possibility would be oversecretion of the cerebrospinal fluid. According to Milhorat,[4] " . . . only obstruction of the cerebrospinal fluid pathways is a proven cause of hydrocephalus."

Cerebrospinal fluid is formed mainly in the choroid plexuses of the lateral ventricles of the brain. There is evidence that a small amount of fluid is added by the choroid plexuses of both the third and fourth ventricles also and that some may be derived from the capillaries in the pia-arachnoid membrane. It passes from the lateral ventricles through the *interventricular foramina (of Monro)* to the third ventricle, through the *cerebral aqueduct* to the fourth ventricle and then out into the subarachnoid space surrounding the medulla via the *lateral foramina (of Luschka).* Some fluid also may pass out of the fourth ventricle via the *foramen of Magendie.*

In the subarachnoid space the cerebrospinal fluid comes to occupy certain dilatations known as the *cisternae* (or cisterns). These are (1) the *cisterna magna,* which lies near the base of the cerebellum where it contacts the medulla; (2) the *cisterna pontis,* which occupies the angle between the pons and the medulla; and (3) the *cisterna basalis,* which lies between the pia mater of the interpeduncular space and the arachnoid anterior to it.

Obstruction of one or more of the communicating foramina

or outlets from the ventricular system results in accumulation of cerebrospinal fluid and expansion of the brain and cranium. If the block occurs in one interventricular foramen, that lateral ventricle becomes distended owing to the continued production of cerebrospinal fluid by the choroid plexus of that ventricle. If it occurs in the cerebral aqueduct, both lateral ventricles and the third ventricle will become distended. If one or more of the outlets from the fourth ventricle become blocked, all aspects of the ventricular system will become distended. In this patient, pressure from the enlarged brain caused a downward excursion of the thin *orbital plates* (i.e., the floors of the anterior cranial fossae and the roofs of the orbits). This, in turn, forced the eyeballs inferiorly, which placed the pupil at about the level of the margin of the inferior eyelid. This is referred to as the "setting sun" sign and is somewhat diagnostic of hydrocephalus in infants. In basic terms, the pressure caused by hydrocephalus results in poor development of the brain, particularly the cerebral cortex.

The cerebrospinal fluid flows around the brain from the cisterna magna and cisterna basalis and is reabsorbed into the venous system via the *arachnoid granulations*. The majority of these lie in the area of the *superior sagittal sinus* (Fig. 8–5).

Fig. 8–5.—Frontal section through the *cerebral hemispheres* and the *superior sagittal sinus*. Note the connections of *emissary veins*, *diploic veins*, *cerebral veins* and *meningeal veins* with the sinus and the *arachnoid granulations* that extend into it.

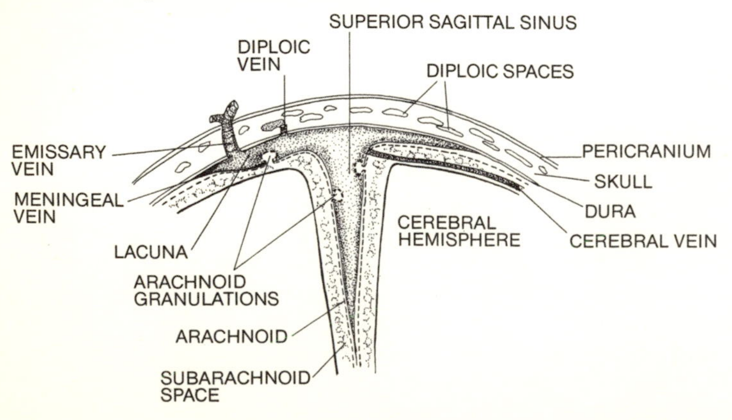

Weed[5] stated that this reabsorptive force was probably ". . . compounded normally of the hydrostatic pressure (subarachnoid pressure minus cerebral venous pressure) and of the colloid osmotic pressure of the blood." Absorption also may occur by the capillaries of the pia-arachnoid membrane or along the cranial and spinal nerves.

CASE 8–6: PROGRESSIVE LEFT-SIDED HEMIPLEGIA

History

A 70-year-old retired businessman with a prolonged history of arteriosclerotic hypertension began to feel lightheaded and dizzy while mowing his lawn. Prior to this time he had not experienced any neurologic signs or headaches. Within a few minutes of the onset of the symptoms, he began to lose consciousness and soon lapsed into a coma. Upon regaining consciousness in the hospital, he demonstrated signs of a progressive left-sided hemiplegia combined with diminution of sensory perception of the trunk, limbs and posterior scalp on that side. There was blood in the cerebrospinal fluid, and the fluid showed increased intracranial pressure. The progression of these deficits reached a plateau and, with time and supportive therapy, his condition improved to the point where he demonstrated a mild paralysis of the limbs on the left side (i.e., a slight, but recognizable, limp) and a barely detectable sensory deficit. He was prescribed antihypertensive drug therapy, freedom from strenuous work (although mild exercise such as walking was encouraged) and a low cholesterol diet. He was free from any remission of the disorder more than 4 years after the initial episode.

Questions

From what did this patient suffer?

Describe the relevant anatomy of this disorder, citing the exact area of the occurrence and the probable blood vessel(s) involved.

Discussion

This man was suffering from a *stroke* (also referred to as *cerebral hemorrhage, cerebrovascular accident* or CVA) which, as the names imply, involves rupture of one of the arteries to the brain. If the hemorrhage is severe enough, the portion of the brain supplied by the ruptured artery will die and the neurologic defects so caused will be permanent. On the other hand, in many cases, as occurred in this one, the defects in motor activity and/or sensory perception seem to abate with time, although usually not totally, after the hemorrhage. In this patient the arteriosclerosis, along with strenuous physical activity and advanced age, apparently combined to cause the stroke.

The lateral surface of a *cerebral hemisphere* (Fig. 8–6) is divided by fissures into a number of lobes. The anteriorly placed *frontal lobe* is separated from the more posteriorly placed *parietal lobe* by the *central fissure*. In front of this fissure is the *anterior central gyrus* and, behind it, the *posterior central* (or *postcentral*) *gyrus*. Brodmann's areas 4 and 6 (which control much of the motor activity of the contralateral side of the body) are located in the anterior central gyrus. The sensory cortex (which perceives sensations derived from the contralateral side of the body) consists of Brodmann's areas 3, 1 and 2 of the postcentral gyrus. The trunk and limbs are represented on the superior aspect of this gyrus extending around the lateral surface of the cerebrum into the *sagittal fissure* between the hemispheres. The face, teeth and tongue have a much larger representation proportionally, occupying most of the lateral and inferior aspects of the postcentral gyrus.

Fig. 8–6. – The lateral surface of the left cerebral hemisphere.

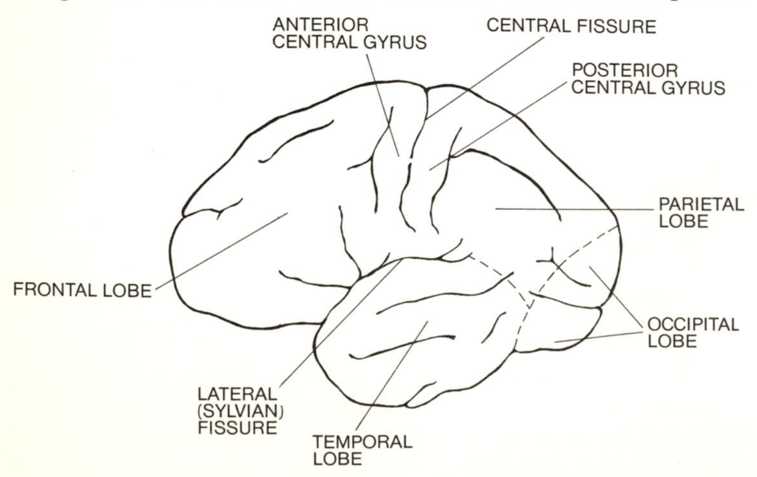

The arterial supply to this area of the brain is derived from the *anterior cerebral arteries*, branches of the *cerebral arterial circle (of Willis)*. Each of these two arteries courses anteriorly, superiorly and then posteriorly in the sagittal fissure between the cerebral hemispheres around the genu of the *corpus callosum*. They supply the frontal lobe, the corpus callosum and the parietal lobe and then terminate by anastomosing with the posterior cerebral arteries. On the medial surface of the cerebral hemisphere, *anterior*, *intermediate* and *posterior medial frontal arteries*—all branches of the anterior cerebral to that side—usually may be found. It most likely was a branch of the posterior medial frontal artery to the right cerebral hemisphere that ruptured in this case, since it supplies the gyri adjacent to the central fissure, especially superiorly. The posterior branches of the intermediate artery may overlap some of the anterior branches of the posterior artery, which might explain the lack of complete hemiplegia following the stroke. The anastomoses with branches of the posterior cerebral artery probably also provide collateral circulation to the area.

CASE 8–7: APHASIA AND ATAXIA

History

A 60-year-old longshoreman had been unable to work for nearly a year and was becoming progressively sicker at the time he visited a physician. He had been suffering from headaches and seizures, and he demonstrated symptoms of aphasia. His fingers, tongue, lips and eyelids twitched almost incessantly, and his speech was frequently slurred. His memory for recent events was defective, and he had demonstrated impaired judgment and manic behavior on numerous occasions. He complained of severe "shooting pains" in the limbs, loss of deep tendon reflexes and impaired deep pain sensations. Most apparent was an ataxic gait and a loss of proprioception. Having the patient stand with the eyes closed, the examining neurologist placed the right upper limb in an abducted position in the horizontal plane and asked the patient to touch the right hand with the left. The man touched the right side near the hip, and then felt up the trunk to the shoulder and out the limb to the hand. He also demonstrated some zones of reduced cutaneous sensation (hypoesthesia) over the thorax and around the nose. A spinal tap revealed slightly increased cerebrospinal fluid pressure, an increase in cell content, elevated total protein and gamma globulin content but normal sugar content. A history revealed that some 14 or 15 years prior to the onset of these symptoms, sores had developed over various parts of the man's body, which were not treated and were very slow in healing.

The patient was presently given massive doses of penicillin (a total of 25 million units delivered at the rate of 1.5 million units per day for over 2 weeks). His condition was apparently too far advanced, however, and he died about 7 months after the start of treatment. An autopsy revealed infiltration of the leptomeninges and cerebral cortex by plasma cells and lymphocytes. A great decrease in the neuronal elements of the cerebral cortex was evident. Similar lesions also were evident in the thalamus, hypothalamus and cerebellum.

Questions

What disease did this man have? Aside from the brain damage that ultimately proved fatal, how do you explain the pain in the limbs and loss of proprioception?

Describe the anatomy relevant to the preceding question.

Discussion

This patient was suffering from *syphilis*, specifically affecting the nervous system (neurosyphilis). It is caused by the spirochete *Treponema pallidum*, which usually is transmitted by direct sexual contact. The disease is commonly described as occurring in three stages. The primary stage or *primary syphilis* is manifested as primary lesions or *chancres*, which are painless sores that appear some 10–40 days after infection. *Secondary syphilis* usually occurs within 3 months and is accompanied by fever, headache and additional skin eruptions. The disease may then become latent, as it did in this patient, reappearing a number of years later as *tertiary syphilis*, in which various organs may be seriously affected.

This patient demonstrated what is known as *taboparesis*, which is a combination of two disease entities caused by the syphilis spirochete. *Tabes dorsalis* is that portion of the overall disease caused by degeneration of the dorsal roots of the spinal nerves. The spirochete causes demyelination of these structures and the dorsal funiculi of the cord, resulting in the cutaneous and proprioceptive disorders. Reflex arcs such as the knee jerk are diminished or abolished, since the propri-

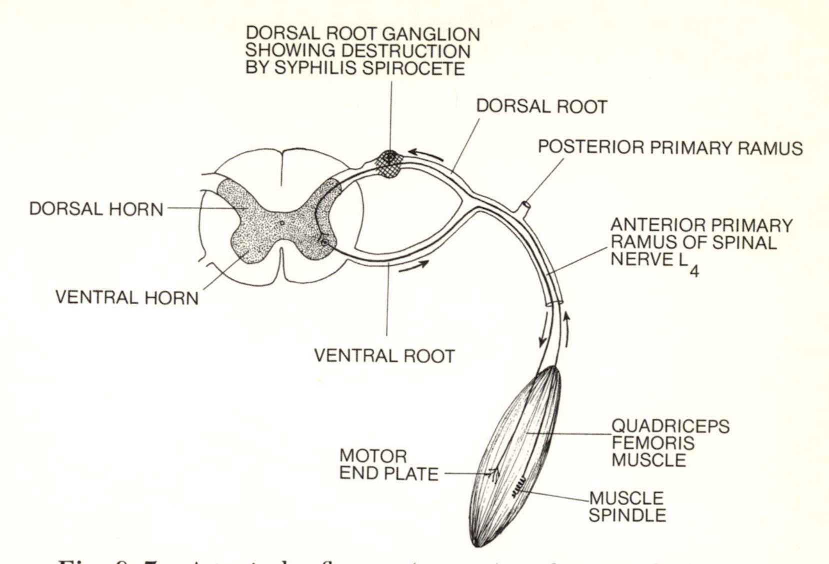

Fig. 8–7.—A typical reflex arc *(arrows)*, in this case the knee jerk. The *cross hatching* represents the *dorsal root ganglion destroyed by the syphilis spirochete.*

oceptive afferent limb is destroyed in tabes dorsalis. Figure 8–7 demonstrates the anatomy of a typical reflex arc and shows how it is affected or destroyed by syphilis. The ataxic gait of the syphilitic patient usually is one of the first symptoms to appear, since the dorsal roots of the lumbar and sacral nerves are the first to be affected by the spirochete.

The *paresis*, manifested as the neurologic symptoms of tremulousness, aphasia, headache and seizures and as memory loss and manic behavior, is the result of *syphilitic meningoencephalitis* caused by the spirochete.

REFERENCES

1. Jackson, C. R. S.: *The Eye in General Practice* (5th ed.; Edinburgh/London: E. & S. Livingstone, Ltd., 1969).
2. Forster, F. M.: *Clinical Neurology* (3d ed.; St. Louis: C. V. Mosby Co., 1973).
3. Barr, M. L.: *The Human Nervous System* (2d ed.; Hagerstown, Md.: Harper & Row, 1974).

4. Milhorat, T. H.: *Hydrocephalus and the Cerebrospinal Fluid* (Baltimore: Williams & Wilkins Co., 1972).
5. Weed, L. H.: Certain anatomical and physiological aspects of the meninges and cerebrospinal fluid, Brain 58:383, 1935.

Chapter 9

ENDOCRINE SYSTEM

CASE 9–1: UPPER TEMPORAL QUADRANT VISUAL LOSS

History

A 57-year-old salesman complained of headache, vomiting, narrowing of the visual field of both eyes, scotomas and a reduction in color vision (mainly for red). A history revealed initial enlargement of the blind spots of both eyes followed by a wedge-shaped loss of vision in the upper temporal quadrants, which gradually enlarged to include the entire quadrant and then spread to the lower quadrant. Ophthalmoscopy revealed a slight degree of papilledema. Mental disturbances and impairment of endocrine function were suspected but not confirmed. Radiographic examination of the skull revealed enlargement of the sella turcica, with atrophy of its floor and erosion and consequent depression of the posterior clinoid processes. Some calcification was evident in the area of the pituitary gland. Radiotherapy was unsuccessful in the treatment of the disorder and therefore surgery was performed. The latter arrested progression of the symptoms.

Questions

What organ was involved in this case and what happened to it?

Describe the anatomy of this organ (from an embryologic, gross, histologic and neurologic aspect).

Discuss its anatomic relationships as they apply to the symptoms described, especially the osteologic and visual changes.

In very general terms describe the functions of this organ.

What hazards might be encountered in its surgical removal?

Discussion

The organ involved in this patient was the *pituitary gland* (hypophysis). The symptoms point directly to a tumor of the gland, most likely in this case a *craniopharyngioma.* According to Martin,[1] pituitary tumors comprise approximately 10% of all cranial neoplasms, with adenomas and craniopharyngiomas the commonest types. Approximately 70% are adenomas, the majority being of the chromophobe type (see below). Only about 2% of pituitary tumors are malignant and they are rarely metastatic. Craniopharyngiomas constitute about 30% of pituitary tumors, are slow growing and demonstrate calcifications in about 75% of patients. They usually occur early in life, although about one fourth of them are found in individuals over 50 years of age.

The pituitary gland contains an *anterior lobe,* the *adenohypophysis,* and a *posterior lobe,* the *neurohypophysis;* these two lobes possess a dual origin. The anterior lobe is derived from a superiorly directed outpocketing of the ectoderm of the mouth of the embryo. This outpocketing is known as *Rathke's pouch.* It pinches off from the oral ectoderm and comes to lie anterior to a conelike depression of the neural ectoderm, the *infundibular process.* The latter is a downgrowth of the diencephalon and will become the neurohypophysis. It remains connected to the undersurface of the brain by the *hypophyseal stalk.* If Rathke's pouch does not close off completely, it may persist as the *craniopharyngeal canal,* and a small amount of pituitary tissue may be found in the posterior wall of the pharynx of an adult.

Both anterior and posterior lobes may be divided into three parts. The adenohypophysis is made up of (1) the *pars tuberalis,* which surrounds the median eminence and upper portion of the neural stalk (see below); (2) the *pars intermedia,* the most posterior part of the adenohypophysis, which comes in contact with the neurohypophysis; and (3) the *pars distalis (pars glandularis),* the anteriormost portion of the adenohypophysis, which comprises the major secretory part of this division of the gland. The neurohypophysis consists of (1) the *median eminence of the tuber cinereum,* which is the superior portion of the hypophyseal stalk at the attachment point to the brain; (2) the *infundibular stem,* the neural portion of the stalk (that portion of the stalk that is surrounded by the pars tuberalis of the anterior lobe); and (3) the *infundibular process (neural lobe, pars nervosa),* which consists of a mass of nerve endings resulting from the downgrowth of neural tissue from the brain. The pars tuberalis, by encircling the median eminence, forms an important relationship between the two portions of the gland, since the anterior lobe lost its anatomical connections in embryogenesis.

The pituitary is supplied by bilateral *superior* and *inferior hypophyseal arteries,* branches of the internal carotid arteries. Interestingly, the pars distalis of the anterior lobe does not receive direct arterial blood but, rather, receives its blood supply from portal veins fed by the superior hypophyseal arteries in the stalk. From this superior plexus of venules or capillaries, blood is conveyed down the stalk to a secondary plexus of sinusoids in the epithelial tissue of the pars distalis. From here, small veins coalesce to drain into the dural sinuses around the pituitary and eventually to the general circulation. In this way, hormones from the adenohypophysis are carried into the blood stream. The posterior lobe has a separate blood supply from the inferior hypophyseal arteries. The nature of the *hypophyseal portal system* has stimulated theories about the association and relationship between the brain and the adenohypophysis. Harris[2] suggested that chemotransmitters liberated by the hypothalamus enter the superior set of capillaries, pass through the portal trunks to the inferior set of sinusoids and thereby influence the adenohypophysis in its secretory function.

The adenohypophysis probably receives sympathetic fibers from the cervical sympathetic chain and parasympathetic fibers from the greater petrosal nerve. The neurohypophysis receives nerve fibers via the *hypothalamo-hypophyseal tract* in the pituitary stalk. This is composed of two divisions: (1) the *supraoptico-hypophyseal tract,* which extends from the supraoptic nucleus in the hypothalamus to the anterior portion of the neurohypophysis; and (2) the *tubero-hypophyseal tract,* which occupies the posterior aspect of the stalk and posterior lobe. Section of these tracts results in atrophy of the neurohypophysis and the onset of diabetes insipidus.

The adenohypophysis contains basically two cell types. The *chromophobes* are agranular and are not thought to secrete hormones. The *chromophils* are granular and are described on the basis of the staining properties of these granules. The granules of *acidophils* stain red with acid-type dyes (e.g., eosin). The *basophil* cell granules stain blue with basic dyes of the methyl blue variety. Of the anterior lobe hormones, it has been suggested, but is as yet unproved, that the acidophils elaborate growth hormone, prolactin, adrenocorticotrophic hormone (ACTH) and thyrotrophic hormone, whereas the basophils secrete the two gonadotrophic hormones. The names of these hormones indicate their functions. How the two cell types produce six hormones is still a question. The posterior lobe contains *pituicytes,* which are specialized neuroglia. The secretions of the posterior lobe are *vasopressin,* the antidiuretic hormone, and *oxytocin (pitocin),* which is responsible for uterine smooth-muscle contraction and for ejection of milk from the breast during lactation. These two hormones are believed to be produced in the hypothalamus and to travel down the axons of the hypothalamo-hypophyseal tract to be stored and eventually secreted by the neurohypophysis.

The pituitary gland is about 8 mm in length, 12 mm in width and 6 mm in height. It is one of the few organs in the body that is heavier in females than in males (approximately 600 mg in females and about 500 mg in males). It is housed within a depression in the body of the sphenoid bone, the *sella turcica* ("Turkish saddle" due to its physical similarity to that structure when viewed from above). Fig. 9–1 demonstrates the relationships of the pituitary gland to the sella turcica and other structures in the midsagittal plane. The student will note the close anatomical relationships to the *sphenoidal air sinus* and the *anterior* and *posterior clinoid processes.* The sella turcica is bridged by the double-layered fold of dura mater, the *diaphragma sellae,* which presents a central hole for transmission of the hypophyseal stalk. Anterior and posterior to the

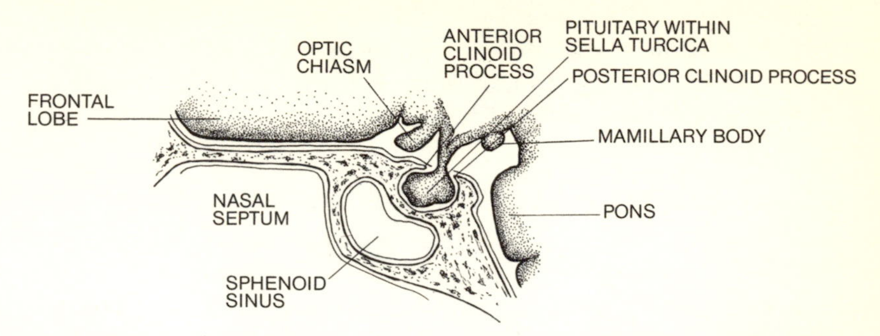

Fig. 9–1.—The *pituitary gland within the sella turcica.* Note the close anatomic association with the *optic chiasm,* a very important relationship in diagnosing pituitary tumors.

stalk, the *intercavernous venous sinuses* communicate between the *cavernous sinuses,* which lie on either side of the sella turcica. Probably the most important relationship is that of the *optic chiasma,* which occupies a position anterosuperior to the pituitary gland. Tumors of the gland frequently cause the following primary physical disturbances: (1) erosion of the sella turcica, as noted in the case history, and (2) pressure on the internal crossing fibers of the optic chiasm. These fibers are derived from connections with photoreceptors that occupy the nasal portion of the retina. This results in a *bitemporal hemianopsia,* in which the visual fields are narrowed bilaterally on their *temporal* aspects.

There are two basic surgical approaches to the pituitary gland. The *frontal approach* involves elevating the frontal lobes of the brain to gain exposure of the gland in the sella turcica behind the optic chiasm. This approach is the most direct and offers the best exposure of the gland, but it tends to traumatize the blood supply to the frontal lobes as well as to tear the delicate filaments of the olfactory nerve as they pass through the cribriform plate of the ethmoid bone (i.e., the olfactory bulb is pulled away with the frontal lobes). These injuries lead to seizures and loss of olfactory sensation. The *intranasal (transsphenoidal) approach* does not disturb the central nervous system and is safer than the frontal approach, but the exposure of the gland is much poorer (i.e., it is basically blind), and the operation may result in sepsis and postoperative rhinorrhea. The frontal approach usually is preferred if the tumor has exceeded the limits of the diaphragma sellae.

CASE 9–2: SKELETAL DECALCIFICATION

History

A 40-year-old woman had complained for some time of muscle weakness, tenderness and pain in the bones, nausea, anorexia, weight loss, polyuria and polydipsia. She had sustained three long bone fractures in the previous 2 years. She was taken to an emergency room with severe abdominal pain, where a physical examination revealed hypercalcemia, hypercalciuria and hypophosphatemia. The serum calcium level was approximately twice normal, and she also had an increased serum phosphatase level. The abdominal pain was traced to a renal calculus that apparently was temporarily lodged in a ureter, causing local spasm, and then was passed, since her discomfort subsided without treatment. Radiologic examination revealed the presence of additional renal calculi, however. Further tests revealed softening of the bones (osteomalacia), some bone cysts and a generalized decalcification of the skeleton. A biopsy was followed by surgery, after which the serum calcium, phosphorus and phosphatase levels returned to within normal limits. The decalcification process was reversed, the bones strengthened and she began to feel well for the first time in years. Her recovery was complete.

Questions

What glands were responsible for the patient's symptoms?

Describe the gross anatomy, embryology, histology and function of these structures.

What was wrong with these glands and what would the surgery basically entail in the attempt to correct the condition?

Discussion

The glands responsible for the symptoms of this patient were the *parathyroids*. The condition is known as *hyperparathyroidism* (also *von Recklinghausen's disease* or *generalized osteitis fibrosa cystica*); it results from a higher than normal degree of activity of these glands. Case 7–3 discusses the etiology and symptoms of *hypoparathyroidism,* which occurred after accidental removal of about half of the normal complement of these glands.

The parathyroid glands usually are light brown, are somewhat oblong to spherical (measuring about 6 mm along their greatest diameter) and weigh somewhere between 30 mg and 60 mg each (total about 250 mg). They are endocrine in nature (see below), their complete removal being incompatible with life. They receive their blood supply from the inferior thyroid arteries, and ligation of these vessels may result in necrosis of the glands. There usually are four parathyroid glands: a *superior set* of two, each of which is found behind the upper pole of the corresponding half of the thyroid gland, and an *inferior set* that is more variable in position. Each possesses its own capsule, although the gland may be imbedded within the substance of the thyroid gland (which also has a capsule), making any but histologic identification difficult to impossible. There may be from two to eight glands; those not in direct association with the thyroid gland may be found elsewhere along the trachea or in the anterior mediastinum in juxtaposition to the thymus gland.

The parathyroids develop during the fifth and sixth weeks of embryogenesis from the entoderm of the *third* and *fourth pharyngeal pouches*. It is interesting to note that the superior set is derived from the fourth pouches, whereas the inferior set is formed, along with the thymus gland, from the third pouches. The superior set tends to maintain a fairly constant position in the adult, but the inferior set migrates inferiorly with the thymus to a variable extent. The glands lose their connections with the pharyngeal pouches at the end of the sixth week of development.

Histologically the glands are composed of two types of cells, the *chief cells,* which do not contain granules, and the *oxyphil cells,* which contain eosinophilic granules. The chief cells are responsible for elaboration of the hormone; the oxyphil cells do not develop until the second decade of life. The two types of cells are arranged in compact strands within the glands.

The parathyroid glands produce *parathyroid hormone* or *parathormone,* which increases the level of serum calcium and promotes the urinary excretion of phosphate. Hyperparathyroidism is characterized by the production of excessive parathormone, which results in the demineralization of bone and the attendant symptoms. This may be caused by hyperplasia of one or more of the glands or by a parathyroid tumor. In either case, the required treatment is surgical, involving either removal of the tumor or excision of a portion of each gland. In the latter event, it is important that the surgeon recognize the potential ectopic nature of the parathyroid glands. The prognosis for recovery usually is good unless the tumor is malignant and metastasis has occurred.

CASE 9–3: ABNORMAL CUTANEOUS PIGMENTATION

History

A 31-year-old woman was observed by her husband to become gradually apathetic about the events of everyday life. She suffered from languor and a feeling of general malaise. Within a few weeks of the onset of these symptoms, pigmented areas began to develop around the face, axillae, groins, external genitalia and areolae of the nipples. These pigmented patches also were visible on the tongue, gums, buccal mucosa and lips. She became progressively weaker, had bouts of diarrhea and vomiting and, as a result, consulted her physician. A physical examination revealed hypotension (100/65), an abnormally small heart with weak pulse and calcifications observable on abdominal X-rays opposite the body of L. V. 1. The serum sodium and chloride levels were lower than normal, the serum potassium elevated. The patient was hypoglycemic. Urinary excretion of 17-ketosteroids was low, and that of 11-oxysteroids absent. An intravenous injection of ACTH did not increase urinary excretion of 17-ketosteroids or 17-hydroxycorticoids. She was treated medically and, with proper management, was able to lead a relatively normal life.

Questions

What is the name of this condition?

What organ, and what specific portion thereof, was responsible for the symptoms described in this case?

Describe the gross anatomy (including blood and nerve supply), embryology and histology of this structure.

Discuss its function, and the rationale for the specific form of treatment prescribed.

Discussion

This condition is known as *Addison's disease*, and it is caused by destruction of the majority of the *cortical portions* of the *suprarenal glands*. These structures also are frequently called *adrenal glands* due to their close relationship to the kidneys (adrenal = "near kidney"), but in the human being they are specifically situated superior to the kidneys, hence the term *suprarenal*. *Adrenalin* was the earlier term for *epinephrine*, which is secreted by the adrenal medulla (see Case 9–4). In animals complete removal of the suprarenal glands results in death in about 15 days. This is traced to a deficiency in the hormones secreted by the adrenal cortex, which are essential to life.

The suprarenal glands are pyramidal-shaped organs, about 5 cm in height, 2.5–3 cm in width and somewhat less than 1 cm in thickness. They weigh about 3–6 gm each and sit astride the superior pole of the kidney, often giving the impression of a "cocked" hat. They consist of an outer *cortex* that surrounds a central *medulla*. They are surrounded by and attached to the renal fascia; they do not descend in company with the kidneys during changes in posture or during respiration. The glands have a profuse arterial blood supply in which a high degree of variation is the rule rather than the exception. Generally a *superior suprarenal artery* is derived from the inferior phrenic artery on each side; this artery may be multiple or absent. A *middle suprarenal artery* is derived from the aorta, and an *inferior suprarenal artery* usually is found emanating from the renal artery on each side. The *suprarenal vein* leaves each gland from its *hilus*. On the right it enters the inferior vena cava directly, whereas on the left it is a tributary of the left renal vein. The nerve supply is derived from the *celiac plexus* and from the *thoracic* and *lumbar splanchnic nerves*. Some of the fibers are postganglionic sympathetic and parasympathetic neurons, but the majority are preganglionic sympathetics that end directly on cells of the suprarenal medulla.

Embryologically the cortex and medulla of this gland are separately derived endocrine organs, which in the human being become intimately associated anatomically during development. The two structures are separate organs in fish. The cortex is derived from mesoderm associated with the urogenital ridge, and it develops earlier than does the medulla. In the newborn infant, a so-called *fetal cortex* exists and, as a result, the glands are relatively extensive at birth, each being one-third the size of the corresponding kidney. The fetal cortex begins to involute after birth so that the absolute weight of each gland decreases. It is replaced by the *adult cortex*, but the gland does not reach its birth weight again until puberty. Because of its developmental relationship with the urogenital ridge, accessory cortical tissue may be found anywhere in the abdomen and pelvis from the level of the kidneys to the scrotum. The medulla arises from neurectoderm; more details of this process will be presented in Case 9–4.

Each suprarenal gland possesses a well-developed fibroelastic *capsule,* which sends *trabeculae* inward toward the medulla. The cortex consists of cords of cells, usually two cells thick, separated from one another by trabeculae and by *sinusoids.* The cortex is morphologically divisible into three relatively distinct zones. The outer *zona glomerulosa* is comparatively thin and contains columnar cells with light-staining cytoplasm, containing some lipid droplets, and a deep-staining nucleus. The intermediate *zona fasciculata* is the widest zone and consists primarily of cords of cuboidal cells. Many lipid droplets may be found in the cytoplasm of the outer two-thirds of this zone. When the lipid is removed in paraffin sections, the cytoplasm of these cells appears empty, and they have been referred to as *clear cells.* The inner *zona reticularis* abuts on the medulla. It is between the other two zones in thickness. The cells contain less lipid than the previously described zones. They possess pyknotic nuclei and an acidophilic cytoplasm and have been termed *compact cells* or *dark cells.*

The suprarenal cortex produces steroid hormones that function importantly in the maintenance of electrolyte balance and in carbohydrate and protein metabolism. On the basis of their biochemical properties, these hormones may be categorized as *mineralocorticoids, glucocorticoids* and *sex hormones.* They are essential to life, as evidenced by the fatal disposition of untreated patients with Addison's disease. Management of this condition consists of substitution replacement therapy with adrenocortical extracts (i.e., cortisone); most of the patient's symptoms will then disappear. Caution must be applied, however, in that abrupt discontinuation of the cortical extract may prove fatal.

CASE 9–4: RECURRING SEVERE HYPERTENSION

History

A 23-year-old man suffered from recurring bouts of severe hypertension during which he experienced a feeling of acute anxiety and the fear of impending death. At the time of these attacks he would sweat profusely, suffer excruciating pounding headaches and experience cardiac arrhythmias (palpitations). Occasionally such symptoms as nausea, vomiting, dizziness, epigastric and/or anginal pain and reduction or temporary loss of sensation in the limbs would be evident. During these attacks his blood pressure would rise to levels as high as 270/140. It could be reduced by adrenolytic substances, e.g., phentolamine. The levels of catecholamines in the urine were elevated considerably above normal. Radiographic examination revealed an abdominal tumor on the left side of the spinal column at about the level of L. V. 1. This was removed surgically, after which the patient's symptoms subsided.

Questions

What is the name of this disorder?

What specific portion of what organ does it affect?

Describe the embryology, histology and function of this organ.

Discuss the rationale for the treatment of choice.

Discussion

This disorder is known as a *pheochromocytoma,* which is a tumor, almost always benign, of the *suprarenal (adrenal) medulla*. These tumors may be solid or cystic, are often partially necrotic and hemorrhagic and may reach a diameter of 10 cm. They cause an active production of the pressor amines (epinephrine and norepinephrine), which are responsible for the hypertension and its attendant symptoms. These substances are found at increased levels in both blood and urine during paroxysmal attacks of hypertension. If the patient is untreated, the attacks may lead to death via cerebrovascular accident or myocardial infarction.

The development of the cortex of the suprarenal gland from mesoderm was discussed in Case 9–3. The medulla has a separate origin from neurectoderm. Cells from this layer invade the medial aspect of the fetal cortex during its development. They stain a yellowish-brown with chrome salts and thus have been termed *chromaffin cells*. The *chromaffin system* consists of these cells of the suprarenal medulla and sympathetic ganglia cells, both of which secrete epinephrine. Indeed, certain investigators have likened the chromaffin cells of the medulla to postganglionic sympathetic neurons, since preganglionic neurons end directly on them.

Medullary cells are polyhedral with a basophilic cytoplasm and a prominent nucleus that contains numerous vesicles. Secretory granules that contain the precursors of epinephrine and norepinephrine are abundant in the cytoplasm. The medullary cells appear to be polarized, one end being adjacent to a capillary, the other to a postcapillary venule. The capillary end contains the nucleus and receives the preganglionic sympathetic fibers, whereas the end abutting on the venule contains the secretory granules and Golgi apparatus.

Epinephrine and norepinephrine are chemically very similar compounds, epinephrine possessing simply one additional methyl group. Epinephrine causes vasoconstriction of the skin and in splanchnic areas but vasodilation in muscle and liver. It produces increased cardiac output while maintaining or slightly elevating blood pressure. Norepinephrine is a peripheral vasoconstrictor, causes increases in both systolic and diastolic blood pressure but has little or no effect on increasing cardiac output. Both substances dilate the coronary arteries and reduce renal blood flow. Epinephrine, therefore, appears to be related to the "fight or flight" mechanism (see also Case 8–1), whereas norephinephrine appears to be the physiologic pressor substance. Epinephrine is the more active (by some 10 times) compound in increasing tissue oxygen consumption and in glycogen mobilization from the liver and glucose uptake by the tissues.

A pheochromocytoma causes overstimulation of the suprarenal medulla and, as a result, the patient experiences symptoms related to the "fight or flight" reaction (anxiety, fear) and those related to elevation of blood pressure. Surgical removal of the tumor diminishes or eliminates these symptoms.

CASE 9–5: ABNORMAL DEPOSITION OF ADIPOSE TISSUE

History

A 28-year-old woman noticed over a period of a few weeks that she was gaining weight and becoming "moon-faced." The majority of the newly acquired adipose tissue was deposited around the neck and trunk, the limbs remaining relatively free of fat (the so-called "buffalo fat" pattern). As the adiposity increased, she noticed curiously that the degree of weight gain did not increase commensurately. The skin became excessively oily, red and mottled. Reddish purple striae (like those seen over the abdomen in pregnancy, the *striae gravidarum)* appeared on the shoulders, breasts, abdominal wall and thighs. Additional hair *(hirsuties)* began to grow on face, trunk and limbs. She was moderately hypertensive, blood pressure ranging from 170/95 to 140/90. Radiographic examination, performed some time after her initial visit, revealed widespread osteoporosis and fractures in two vertebrae. Renal calculi were present bilaterally. Tests did not reveal a diabetic condition. There was loss of libido combined with amenorrhea. Sophisticated steroid metabolism tests revealed an increased level of urinary 11-hydroxycorticoids, but the urinary 17-oxosteroid level was within normal limits. She was treated with both irradiation and surgery. She improved briefly after treatment, but her condition steadily deteriorated after that and she died 4 years after diagnosis.

Questions

What is the name of this disease and what organ or organs are responsible for its manifestation?

Why did the woman's weight gain not reflect the degree of addition of adipose tissue?

What was the cause of the osteoporosis and the renal calculi?

Discuss the rationale for treatment.

Generally, what do you think is the prognosis for this disease?

Discussion

This patient was suffering from the *Cushing disease* or *syndrome,* the symptoms of which are generally agreed to be due to overactivity of the *suprarenal cortex.* Whether this, in turn, is due to a direct problem, i.e., a tumor or hyperplasia of the suprarenal cortex itself, or due to a similar, but indirect, problem of the pituitary gland (which controls the activity of the cortex through production of ACTH) is still not certain. The Cushing disease occurs in association with basophil adenomas of the pituitary, less frequently with oat-cell carcinomas of a bronchus and even less frequently with tumors of other organs. The fact that tumors are not always identifiable in this disease and that it may be at least partially treated by subtotal adrenalectomy has led some investigators to believe that the primary problem lies within the suprarenal cortex. This organ was described in Case 9 – 3, and the pituitary in Case 9 – 1.

The Cushing disease affects normal gluconeogenesis due to the abnormally high level of circulating corticosteroids. As a result, protein is converted to carbohydrate and then stored as fat. Thus, the patient is producing and storing fat at the expense of muscle, and weight change may be minimal even though the individual is "getting fat." The osteoporosis also is caused by the breakdown of protein, in this case from the bone matrix. Although there is no defect in calcium metabolism associated with this syndrome, calcium is released from the bone as the protein is broken down. This results in increased urinary excretion of calcium and the possible build-up of calcium within the kidneys, manifested as calculi or stones.

The patient was treated by irradiation of the pituitary gland combined with subtotal adrenalectomy in order to (1) arrest a pituitary tumor, if present, and (2) treat the problem directly by reducing the cortical hyperplasia. The prognosis for survival, however, is poor, since the underlying cause of the cortical hyperplasia frequently is obscure and remains after the subtotal adrenalectomy.

CASE 9–6: ABNORMAL GLUCOSE TOLERANCE

History

A 41-year-old man with an obesity problem made an appointment for a physical examination. As a part of this examination the physician included a glucose tolerance test. Blood sugar levels were recorded during fasting and at 30, 60, 90 and 120 minutes after administration of 100 gm of carbohydrate. The venous blood sugar readings (mg of true glucose/100 ml blood) were as follows: fasting, greater than 110; at 30 minutes after carbohydrate administration, greater than 160; at 60 minutes, greater than 160; at 90 minutes, greater than 140 and at 120 minutes, greater than 120. The patient was instructed to adhere to a strict diet and, as a result, his condition became manageable and stable.

Questions

These glucose tolerance test figures are diagnostic of what disease?

What portion of what organ is responsible for this metabolic deficiency?

Describe the embryology, gross anatomy (including blood supply, lymphatic drainage and innervation), histology and function of this organ.

Discussion

These test figures are characteristic of (i.e., diagnostic for) *diabetes mellitus,* in which there is a deficiency in *insulin* production by the *pancreatic islets (of Langerhans).* Insulin is very important in carbohydrate metabolism, causing an increase in the uptake of glucose by the cells of the body and in the conversion of glucose to glycogen. Disturbance of this function results in increased blood sugar levels, as observed in this patient. The normal figures for this test are as follows: fasting, less than 100; 30 minutes, less than 160; 60 minutes, less than 160 and 120 minutes, less than 100.

The pancreas develops from *dorsal* and *ventral primordia* or *buds*. The dorsal bud contributes all of the neck, body and tail of the organ, plus part of the head and the *accessory pancreatic duct (of Santorini).* The ventral bud forms the remainder of

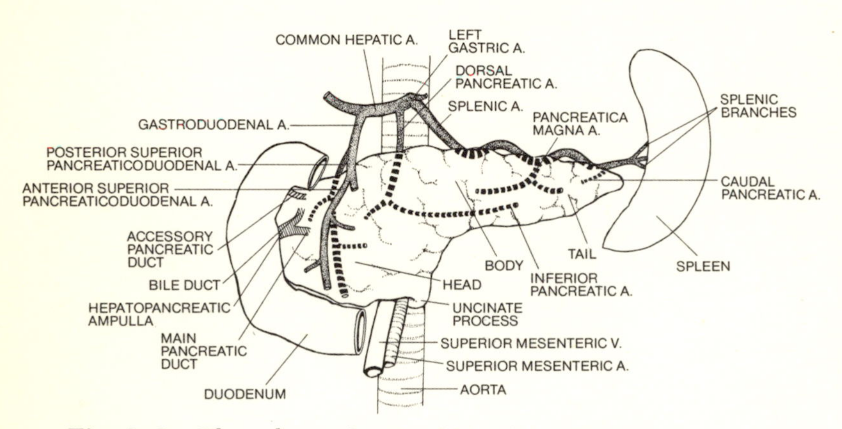

Fig. 9–2.—The relationships and blood supply of the pancreas.

the head and the *main pancreatic duct (of Wirsung).* In the adult the pancreas becomes largely retroperitoneal.

The *head* of the pancreas lies within the horseshoe-shaped contour of the duodenum (Fig. 9–2), into which its ducts empty. The head is joined to the *body* by a short *neck*. The *tail* projects to the left in the lienorenal ligament to come into close association with the spleen. The *uncinate process* projects from the inferior aspect of the head and gives the impression of being notched by the *superior mesenteric artery* and *vein* as they emerge from behind the gland. The pancreas is supplied mainly by the *splenic artery,* which runs a tortuous course across its superior border. In this course the splenic artery gives off the *dorsal pancreatic, pancreatica magna* and *caudal pancreatic arteries.* The gland also receives blood from the *gastroduodenal artery (anterior* and *posterior superior pancreaticoduodenal branches)* and the *superior mesenteric artery (anterior* and *posterior inferior pancreaticoduodenal branches).* The venous drainage is more variable but generally follows the arteries, the blood draining eventually into *portal, splenic* and *superior mesenteric veins.* Autonomic nerves reach the pancreas through the *celiac* and *superior mesenteric plexuses;* sympathetic fibers are derived from the thoracic splanchnic nerves and parasympathetic fibers are derived from the vagus nerve. Lamellated (pacinian) corpuscles are numerous in the pancreas, but their function is unknown. The parasympathetic fibers probably stimulate secretions. Humoral factors, of course, also play a large part in this function. Lymphatic vessels drain to adjacent lymph nodes (splenic, mesenteric, gastric, hepatic, celiac).

The pancreas is both endocrine and exocrine in function. The *exocrine portion* consists of a compound acinous gland, producing zymogen granules and secreting their products through ducts. The *endocrine portion* comprises the *islets of Langerhans.* These consist of three types of polyhedral cells, one surface of each abutting on a capillary. The capillaries possess a fenestrated endothelium for facilitation of hormone passage from the islet cell. The *alpha cells* comprise 15–20% of the total endocrine component. Their granules stain yellow to orange to red with Mallory-azan or aldehyde fuchsin-negative stains. They secrete glucagon, which is antagonistic to insulin. The *beta cells* account for 75–80% of the pancreatic islet cells. Their granules stain a brownish orange with aldehyde fuchsin-positive; these are the cells that secrete insulin. *Delta cells* comprise approximately 5% of the endocrine component. They contain blue-staining granules.

REFERENCES

1. Martin, L.: *Clinical Endocrinology* (5th ed.; London: J. & A. Churchill, Ltd., 1969).
2. Harris, G. W.: Neural control of the pituitary gland, Br. Med. J. 2:559, 1951.

Chapter 10

REPRODUCTIVE SYSTEM

CASE 10–1: PUDENDAL NERVE BLOCK

History

A 22-year-old woman, experiencing her first pregnancy, had desired with the concurrence of her husband to undergo natural childbirth without the use of any anesthetics. They spent a number of hours reading on the subject and attending seminars and study groups. As she entered labor, however, the pain became severe and her obstetrician, fearing that delivery might be complicated under these circumstances, ordered a bilateral pudendal nerve block. The baby was born shortly thereafter in what was a perfectly normal delivery.

Questions

Describe the boundaries and subdivisions of the female perineum.

Briefly discuss the lymphatic drainage of the pelvis and perineum.

Describe the origin, course and distribution of the pudendal nerve.

How do you think this nerve should be approached in anesthetizing it?

What is the main landmark used in identifying it?

What other nerve(s) need to be deadened in order to insure complete anesthesia of the perineal area?

Discussion

The *perineum* is the diamond-shaped area located between the thighs at their attachment to the trunk. Its anterior limit is the *pubic symphysis*, its two lateral limits the *ischial tuberosities;* posteriorly it is limited by the tip of the *coccyx* (Fig.

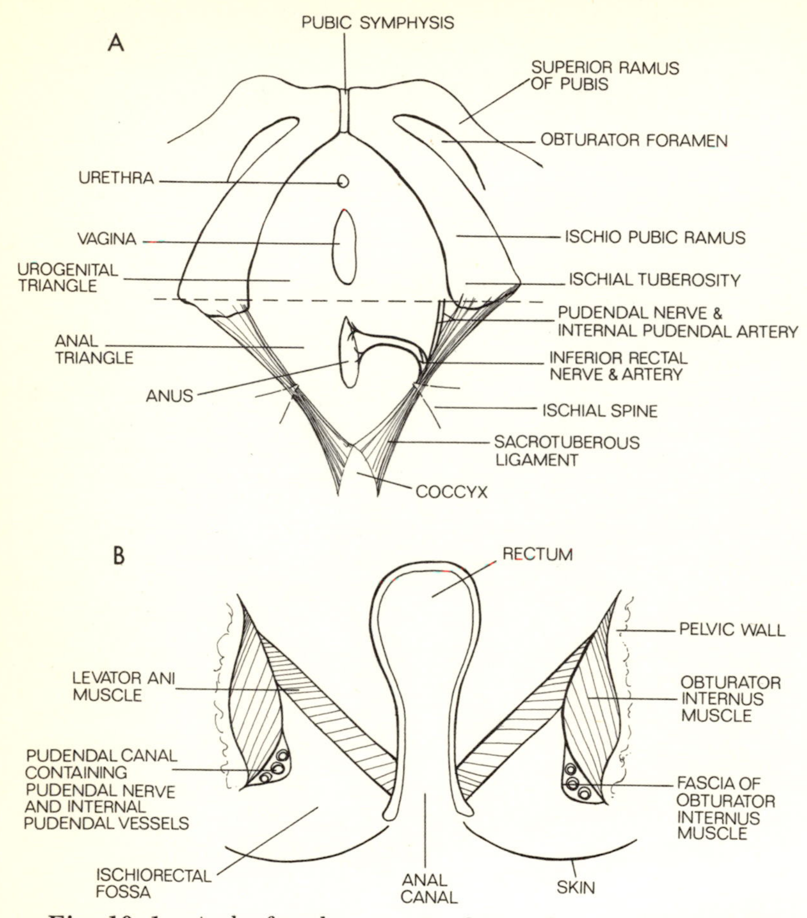

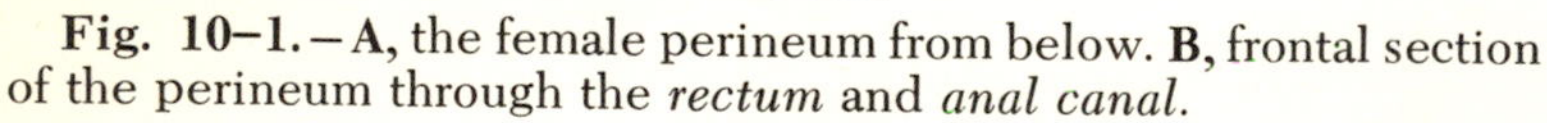
Fig. 10–1.—**A,** the female perineum from below. **B,** frontal section of the perineum through the *rectum* and *anal canal.*

10–1, A). Between the pubic symphysis and the ischial tuberosities extend the bony *ischiopubic rami.* The angle formed by these structures is the *pubic arch.* The rhomboid is completed by the sacrotuberous ligaments, which extend from the sacrum and coccyx to the ischial tuberosities. These ligaments overlie the ischial spines and associated sacrospinous ligaments. The gluteus maximus muscles in turn overlie the sacrotuberous ligaments. An imaginary line (see broken line, Fig.

10–1, A) drawn from one ischial tuberosity to the other divides the diamond-shaped area into two triangles, an anterior *urogenital triangle* and a posterior *anal triangle*. The former, which corresponds to the position and extent of the muscular *urogenital diaphragm* (composed of the *deep transverse perineal* and *sphincter urethrae muscles*), contains basically the *root of the clitoris* and the urethral and *vaginal openings*. The urogenital area is covered superficially by skin and subcutaneous tissue comprising the *labia major* and *labia minor*.

The anal triangle contains the terminal *anal canal* and on either side of this the two *ischiorectal fossae* (Fig. 10–1, B). The latter are bounded medially by the anal canal and its surrounding *external anal sphincter muscle;* superomedially by the *pelvic diaphragm*, which is composed of the *levator ani* and *coccygeus muscles;* laterally by the fascia over the *obturator internus muscle;* and inferiorly by skin. The ischiorectal fossae are fat-filled compartments traversed by branches of the *inferior rectal artery* and *nerve*. Because of their fat content, they are very distensible, allowing expansion of the rectum and anus during defecation. They are sometimes the seat of infectious processes emanating from various anal or perianal disorders and, because of their distensibility, may allow the accumulation of considerable quantities of pus. Between the anus and the vagina in the midline is the *perineal body* or *central tendon of the perineum*, a fibromuscular tissue mass that is very important to the integrity and support of the perineum, especially in the female. This structure is incised in an episiotomy to prevent tearing and subsequent scar formation during parturition. Repeated tearing and scarring frequently leads to a lack of support of the female pelvic floor (see also Case 10–7).

A word about the lymphatic drainage of the pelvis and perineum is in order at this point due to its importance in tracing the spread of metastatic or infectious processes from these areas. In very general terms, the pelvic viscera in both sexes and the testis in the male drain to internal iliac nodes, thence to common iliac nodes and eventually to lumbar nodes. The terminal anal canal and adjacent skin and subcutaneous tissue of the perineum drain to superficial inguinal lymph nodes.

The glans of the penis and clitoris drain to deep inguinal nodes. Both groups of inguinal nodes send their efferents to external iliac nodes, the latter then draining to common iliac and eventually to lumbar nodes. Thus, although the final destination is the same (i.e., the lumbar lymph trunks and eventually the cisterna chyli and thoracic duct), the specific draining nodes differ for various areas of the perineum. A knowledge of these pathways is important in the early detection and possible eradication of metastatic disease.

The *pudendal nerve* is a branch of the sacral plexus (anterior primary rami of S_{2-4}) emanating within the pelvis. It is both motor and sensory to the perineum and also carries postganglionic sympathetic fibers to this region. It follows an interesting course, passing out of the pelvis through the greater sciatic foramen inferior to the piriformis muscle, only to immediately enter the perineum through the lesser sciatic notch after crossing the external surface of the ischial spine. Immediately upon entering the perineum, it comes to lie on the lateral wall of the ischiorectal fossa within an envelope, so to speak, of the obturator internus fascia (see Fig. 10–1, B). This is known as the *pudendal (Alcock's) canal.* The nerve proceeds in an anterior direction in this canal in company with the *internal pudendal artery* and *vein.* The following are the branches of the pudendal nerve: (1) The *inferior rectal nerve,* which pierces the medial wall of the pudendal canal; traverses the ischiorectal fossa (see above); and supplies the external anal sphincter, the anal canal up to the pectinate line and the skin around the anus. (2) The *perineal nerve,* which with its superficial and deep branches courses anteriorly into the urogenital triangle and supplies the *bulbospongiosus, ischiocavernosus* and *superficial transverse perineal muscles* and the labia in the female and scrotum in the male (*posterior labial* and *scrotal nerves*). (3) *The dorsal nerve of the clitoris* (or *penis in the male*), which supplies the muscles of the urogenital diaphragm (deep transverse perineal and sphincter urethrae) and the *prepuce and glans of the clitoris* (or *penis*) and associated skin.

The pudendal nerve block is the safest type of anesthesia for the obstetrician to use.[1] The *ischial spine* is the important landmark in locating this nerve. The following excerpts on pudendal nerve block have been taken from Tenney and Little's[1] textbook on *Clinical Obstetrics:* "A small wheal of anesthetic is placed about 1½–2 inches from the anus at the 10 and 2 o'clock positions. A finger is placed on the ischial spine, and the needle is inserted in the direction of the tip of the finger on the spine. Here, on withdrawing the barrel of the needle, a little blood may be obtained if the pudendal vessels have been entered. The pudendal nerve is close by, lying medial to and behind the ischial spine. As the sacrospinous ligament is pierced, a small dose of about 2 cc. (xylocaine) is administered. Once through the ligament, the syringe is again withdrawn to be sure that no vessel has been penetrated. Then about 8 to 10 cc. is injected. The syringe is replenished and another 10 cc. is injected as the needle is gradually withdrawn through the skin. This procedure is followed on both sides."

The perineum also is supplied by *anterior labial* (*anterior scrotal* in the male) *branches* of the *ilioinguinal nerve.* This is a branch of the anterior primary ramus of L_1, which follows a course between the muscular layers of the anterolateral abdominal wall to the iliac crest and then accompanies the *round ligament of the uterus* (*spermatic cord* in the male) through the inguinal canal. After emerging from the superficial inguinal ring, it supplies the anterior portion of the labia in the female and adjacent skin of the thigh. Thus, according to Tenney and Little,[1] ". . . a second infiltration with 10 cc. laterally and anteriorly up through the labia majora is very helpful."

CASE 10–2: SEX CHROMOSOME ABNORMALITY IN A MALE

History

A 27-year-old man, married for 8 years but childless, made an appointment with a local physician in an attempt to determine if he was sterile and, if so, why. Prior to this time, he could remember only two or three doctor's appointments in his life, all for minor ills, and he could not recall ever undergoing a physical examination. A complete physical examination at this time revealed the following immediately obvious disorders: (1) *microrchidia* (the testes, although descended, were very small and firm); (2) slight *gynecomastia* (development of the male mammary gland); and (3) reduction in the normal length of the penis, a small, taut scrotum and scanty pubic hair, more female than male in type. The patient needed to shave only infrequently. Laboratory data revealed a marked increase in urinary gonadotrophin excretion and a significant reduction of serum androgens. Examination of a semen specimen revealed that it was void of sperm. Buccal smears were sex chromatin positive and karyotypes demonstrated an extra chromosome. A history revealed consistently poor intellectual performance in schooling and various occupations, and an intelligence test showed the patient to be slightly retarded mentally.

Questions

What is the name of this condition (syndrome) and what is its cause?

What are some of the recent breakthroughs in cytogenetics that have allowed an understanding of such disorders?

Review briefly the significant details of mitosis and meiosis and discuss how cells can end up with too many or too few chromosomes.

What determines "maleness" or "femaleness" in an individual?

What is the significance of sex chromatin-positive cells in males?

What other anatomical or clinical abnormalities would you expect to observe as a result of this condition?

What would you estimate the frequency of this disorder to be (per number of male births)?

Discussion

This condition is known as the *Klinefelter syndrome* and its cause is *nondisjunction* of an X chromosome so that the individual so afflicted possesses an XXY sex chromosomal complement.

Before going into a description of how this and related conditions actually come about, it is important to briefly review some of the major technical advances in cytogenetics that have allowed investigators to understand the causes of such conditions and to begin to develop technics designed to reduce their incidence. It is interesting to note that many of these findings were "accidental," coming about as a byproduct of a different investigation. The first of these, reported by Barr and Bertram[2] in 1949, was that interphase nuclei of neurons from female cats possessed a heterochromatic body, whereas those from male cats did not. These structures, originally called Barr bodies and now known as sex chromatin bodies, represent the inactive X chromosome of female somatic cells and provide a simple morphologic means of distinguishing male from female cells. The second such discovery, made by Hsu and Pomerat[3] in 1953, was that hypotonic solutions cause cell nuclei to swell, accompanied by a disruption of the mitotic apparatus. As a result, the chromosomes of metaphase cells are dispersed, and it was only after this knowledge was put to practical use that the human diploid number was correctly reported (1956) as 46.[4] A third major technical breakthrough was that reported by Nowell[5] in 1960 that phytohemagglutinin, an extract from the bean, *Phaseolus vulgaris,* was not only an efficient red blood cell agglutinin but that, if allowed to remain in culture media, was mitogenic (causing to undergo mitosis) to mammalian lymphocytes. As a result, it became a simple matter to obtain cell samples (from peripheral blood of patients) for chromosome analysis. Prior to this discovery, cells had to be obtained from testicular biopsies or from various somatic tissues, a procedure that frequently caused discomfort to the patient. Other agents or technics that have been useful in the study of chromosome disorders have been (1) colchicine, which arrests and therefore accumulates cells in metaphase, thus increasing the yield of mitotic spreads; (2) ^{3}H-thymidine, a radioisotope specific for newly replicated DNA, which, on the basis of early or late replication, allows investigators to determine active from inactive chromosomes (or segments of chromosomes), respectively; and (3) various staining procedures such as Feulgen, fluorescent dyes (e.g., quinacrine mustard) and Giemsa, which assist in identifying DNA in its various functional states.

In very basic terms, *mitosis* consists of the doubling of each chromosome in such a way that the newly formed chromosome is an exact replica of the one originally present. In the process of cell division, each daughter cell receives an identical set and number (the *diploid* number, or 2n), thereby assuring that there will be no significant differences between the two (unless of course, one is affected variously through environmental stimuli or mutation). *Meiosis*, on the other hand, a more complex procedure, consists of two divisions. The first, often called the *reduction* division, results in each cell having *half* the normal diploid number of chromosomes (i.e., the *haploid* number, or n), one of each pair. The second meiotic division, which is essentially mitotic in nature and referred to as *equational*, consists of a duplication and separation of the haploid set so that four cells, each with half the original chromosome complement, are derived from the original diploid germ cell. During the process there is extensive *crossing over* (exchange) between chromosomes so that none of the final products (gametes) is the same as any of the others. This phenomenon provides for individual genetic differences. The union of one such cell from male and female restores the diploid number, and a new individual is created. If one or more chromosomes (either autosomes or sex chromosomes) should fail to separate normally during cell division, the condition is referred to as *nondisjunction*, and the result is *aneuploidy* (an abnormal number of chromosomes in a given cell). Nondisjunction can occur in both mitosis and meiosis. An example of human nondisjunction with viable clinical demonstrability is when the gamete from the female contains two Xs rather than the normal one and unites with that from the male containing a single Y chromosome (normal). This probably occurs during meiosis in this case. The result is a zygote with 47 chromosomes rather than the normal 46 and with an XXY sex chromosome constitution. This is an example of the Klinefelter syndrome, as described in this case history; other, less frequently occurring examples of this syndrome may contain an XXYY, XXXY, XXXYY or even an XXXXY constitution.

In general, it can be stated that an individual is a "male" if he possesses at least one Y chromosome and a "female" if she possesses no Y chromosomes, regardless of the nature of the remainder of the sex chromosome complement. As a rule, the more aberrant the sex chromosome constitution, the more unusual the phenotypic characteristics, both sexual and somatic. The student also should be aware that *mosaicism,* in which different tissues or regions of the body may possess different sex chromosomal complements, is frequently observed in individuals with cytogenetic disorders. Thus, an individual may have a normal blood lymphocyte chromosome constitution while demonstrating sex chromosomal aneuploidy elsewhere.

Lyon[6] in 1961 suggested that the heterochromatic chromosome of the female mouse somatic interphase nuclei was an X chromosome of either maternal or paternal origin, and that it was genetically inactive (this is now known as the famous Lyon hypothesis). The prerequisite for this hypothesis was the assumption that the Barr body or sex chromatin body represented the interphase condensation of one female X chromosome. Evidence for this assumption, which is now accepted by most investigators in the field, has consisted of the following: (1) sex chromatin almost always is limited to nuclei with more than one X chromosome, i.e., it is not evident in normal male cells (the exception to this rule is in certain animals, notably marsupials, in which male cells frequently demonstrate sex chromatin); (2) XO female mice are phenotypically normal; (3) in normal female somatic cells, one X is late replicating with regard to its DNA, a condition strongly associated with genetic inactivity; (4) heterochromatin synthesizes RNA at a lower rate than does euchromatin (i.e., heterochromatin is genetically *less* active than euchromatin) and (5) when more than two X chromosomes are present (i.e., aneuploidy), the number of sex chromatin bodies is one less than the total number of X chromosomes present, and all but one are late replicating. Therefore, in the patient described here, the fact that a single sex chromatin body was observed in cells taken from a buccal smear (somatic cells) strongly indicated that the patient had two X chromosomes instead of one. Since he possessed a male phenotype, the physician was fairly sure of an XXY sex chromosome constitution, i.e., the Klinefelter syndrome. This was confirmed by preparing a karyotype of the patient's chromosomes (Fig. 10–2).

Other clinical symptoms related to this condition that might aisle, tumbled down into the cracks where the floorboards had rusted, where the cold winter air crept up, where the salt from the roadways ate into the very structure of everything and decayed it bit by bit.

Monica's mother slammed the passenger door.

A silver locket swung from the rear-view mirror. Just like their hearts, her father's image was hidden inside.

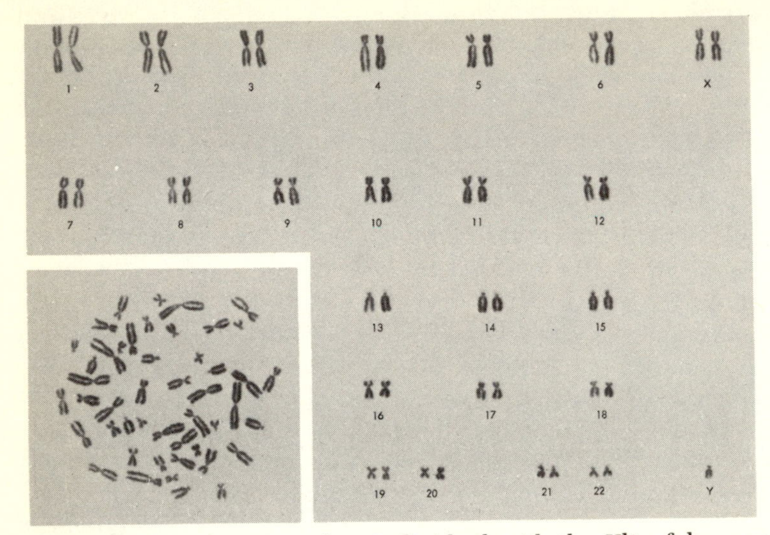

Fig. 10–2. – Karyotype of an individual with the Klinefelter syndrome (courtesy of J. L. German, III, M.D., The New York Blood Center).

be detected during surgery, biopsy or autopsy, or as a result of specialized examinations, are as follows: small epididymides, prostate gland and seminal vesicles; high-pitched voice, precocious osteoporosis; certain psychopathologic traits; pathologic changes in the seminiferous tubules; degeneration of the Leydig cells and scant or absent spermatogenesis.

This disorder occurs in about one in 400 male births and is therefore one of the more frequently observed cytogenetic abnormalities. Behavioral studies have indicated that males with an extra X chromosome, and especially with an extra Y, are more frequently represented in mental institution and prison populations than are normals. The reasons for this are unclear. Although various studies and numerous data suggest that these extra chromosomes are inactive, the rather consistent observation of phenotypic abnormalities in patients who possess them suggests that, at some point in the life of the individual, such chromosomes played a significant role. Certain investigators are currently of the opinion that not all genes on a so-called "inactive" chromosome are indeed inactive, and that some may remain active beyond embryonic development.

CASE 10–3: SEX CHROMOSOME ABNORMALITY IN A FEMALE

History

An anxious mother brought her 16-year-old daughter to the family physician, prompted by the observation that the girl had apparently passed through puberty without undergoing any of the external indications of this event. For example, the girl had never menstruated *(primary amenorrhea)*, absence of breast development was accompanied by an infantile appearance of the external genitalia and she had scant axillary and pubic hair. The mother revealed to the physician that her daughter had always seemed a little subnormal mentally and that of late she had demonstrated increasingly strong inhibitions, an absence of aggressive behavior, a lack of initiative and little or no interest in individuals of the opposite sex. A physical examination showed that the girl was short in stature (the breakoff point between normal and subnormal height usually is considered to be 150 cm, or 59 inches). In addition, she had the following physical characteristics: a slightly webbed neck with a low hairline, a shield-like chest with widely spaced nipples, multiple pigmented nevi, short fourth metacarpal bones and hypoplastic fingernails. Laboratory tests revealed a low serum estrogen level and a low 17-ketosteroid urinary level. Cells from buccal smears were sex chromatin negative.

Questions

What is this condition (syndrome) known as, and what is its cause?

What are some other physical characteristics associated with this disorder that you might expect to observe?

What would you estimate the incidence of this syndrome to be (in numbers of female births)?

Discussion

The symptoms described in this case are characteristic of the *Turner syndrome* (the term *gonadal dysgenesis* is used interchangeably to identify this condition). It is caused by the absence of one X chromosome, i.e., the chromosome constitution is 45, XO. It is believed that this results from the loss of either the maternal or paternal X chromosome through nondisjunction during gametogenesis (in one sex or the other) or during the first cleavage division. For a discussion of the basic concepts of mitosis, meiosis and nondisjunction and the significance of sex chromatin, see Case 10–2. Individuals afflicted with the Turner syndrome are females. It is interesting to note that, although the XO condition in human beings is viable, the YO sex chromosome constitution is not. Another interesting observation is that in contrast to the human, XO female mice are phenotypically normal.[6] Figure 10–3 is a karyotype of an individual with the Turner syndrome.

Other anatomical or clinical findings in individuals with this chromosome make-up are a lack of germ cells in the gonads, infantile internal genitalia, atrophy of the vaginal epithelium and sterility. Coarctation of the aorta, webbing of the fingers and/or toes, epicanthic folds, bilateral cataracts, partial or total deafness, idiopathic hypertension and renal anomalies also are sometimes observed. The Turner syndrome is found in

Fig. 10–3. – Karyotype of an individual with the Turner syndrome (courtesy of J. L. German, III, M. D., The New York Blood Center).

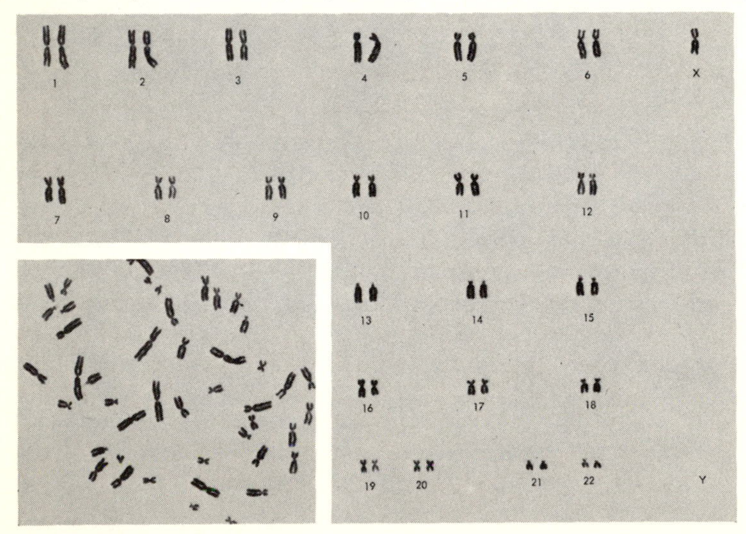

approximately one in 3,000 female births, and is therefore relatively infrequent compared to certain other chromosomal abnormalities.

CASE 10–4: AUTOSOME ABNORMALITY IN A NEWBORN INFANT

History

A newborn female child demonstrated hypotonia (reduced muscle tone), oblique palpebral fissures, pronounced epicanthal folds and short broad hands. Motor development of the child was retarded in that she could sit up only by the age of 12 months and was not able to walk until she was 3 years old. As development proceeded, severe mental retardation was observed (at 6 years of age, the child's IQ was 45), and the following additional symptoms were observed; a simian crease (four-finger line) on the hands; short, crooked fifth fingers; hyperflexibility of the joints; protrusion and furrowing of the tongue; irregular and abnormal teeth; flattened occiput and a narrow, elevated palate. Cytogenetic investigations revealed normal sex chromatin, but one extra chromosome (i.e., a total of 47 chromosomes). Laboratory tests revealed increased alkaline phosphatase and galactose-1-phosphate-uridyl-transferase activity in polymorphonuclear leukocytes, a reduction of various biochemical compounds in the urine and increases in blood calcium and serum uric acid levels, among others.

Questions

What is the name of this condition (syndrome) and what is its cause?

How do autosomal abnormalities compare in severity of phenotypic disorders with those of the sex chromosomes?

What are some other clinical or anatomic findings that might be associated with this condition?

Does the age of the parents have any bearing on the incidence of this disorder and, if so, what might be some reasons for this?

Discuss some of the causes of chromosome breakage, clumping, stickiness, etc., that might result in clinical problems.

Could the condition described above ever be observed in an individual with the normal number of chromosomes?

How can our knowledge of cytogenetics be applied to help combat disorders of this nature?

Discussion

This condition is known as the *Down syndrome* (synonyms are *mongolism* and *trisomy 21*), and its cause is an extra chromosome of the G series, specifically autosome 21. A karyotype of this condition is presented in Figure 10–4, A. This is probably the best known of the chromosome disorders in that practically all, if not all, individuals have witnessed a person with this disorder. The oblique slant of the palpebral fissures and the prominent epicanthal folds are responsible for the term "mongolism." This condition may be caused by nondisjunction during either meiosis or mitosis (first cleavage division of the zygote). Most investigators currently believe that it occurs as a result of meiotic nondisjunction in the ovum (for a discussion of mitosis, meiosis and nondisjunction, see Case 10–2).

In general, it can be stated that the loss of a chromosome is more severe than the addition of one (there are no viable autosomal monosomies, and the only viable sex chromosomal monosomy is XO, the Turner syndrome). In addition, autosomal trisomies are more severe than those involving the sex chromosomes. Other autosomal trisomies that have been reported are of the no. 18 chromosome (Edwards) and 13 chromosome (Patau). In both of these there are multiple defects, e.g., harelip, cleft palate and heart disorders, and individuals so afflicted seldom live more than a few months. In addition, it should be pointed out that there are autosomal abnormalities other than aneuploidy, e.g., *deletions* of small pieces of chromosomes. This occurs in the *cri du chat syndrome,* in which a small piece of one of the no. 5 chromosomes is missing. This small deficiency results in the child possessing a small round face, psychomotor retardation and the cry of a distressed seagull or kitten (hence the name "cri du chat"). A deletion of a small portion of autosome no. 4 results in even more severe symptoms.

Other physical symptoms that may be apparent in the Down syndrome are strabismus; nystagmus; umbilical hernia; cryptorchidism; ventricular septal defects; brain disorders; abnor-

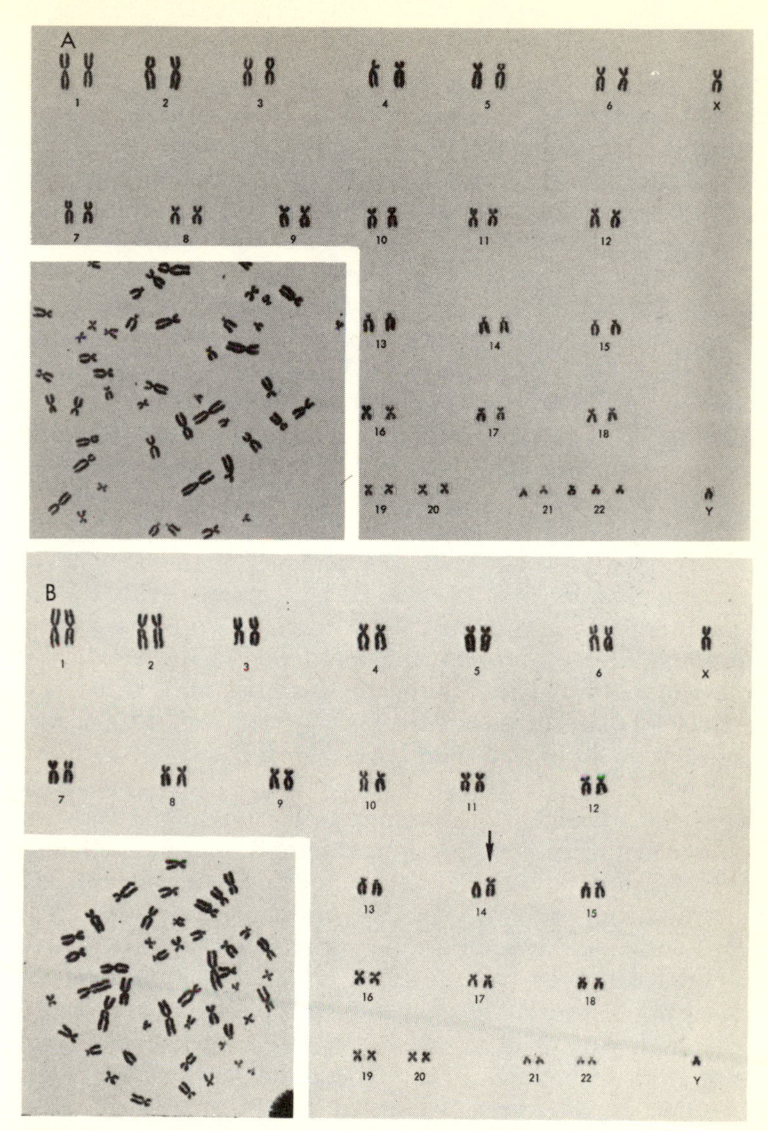

Fig. 10–4.—**A,** karyotype of an individual with the Down syndrome, or mongolism. **B,** karyotype of an individual with the normal diploid number of chromosomes (46) but with the Down syndrome. Note the translocation *(arrow)* of a no. 21 autosome to a no. 14 autosome, thereby constituting a trisomy 21. (Courtesy of J. L. German, III, M. D., The New York Blood Center; reprinted by permission, *American Scientist,* Journal of Sigma Xi, The Scientific Research Society of North America.)

mal thyroid, adrenal and thymus glands; megacolon; microcolon and disorders of ovarian function. The incidence of acute leukemia is three to 15 times higher in children with the Down syndrome than in those in the general population, and an association of this disease with Down patients of all ages has been observed. Only about 4% of all individuals with the syndrome can read with any comprehension, and only approximately 2% can write. Children afflicted with the Down syndrome often are cheerful and happy, although they also may be stubborn, and they frequently possess a capacity for mimicking other children and adults. According to Bartalos and Baramki,[7] "They enjoy dancing, singing and listening to music and are good-natured and sociable." It is generally conceded that the life span of Down patients is considerably less than that of the general population.

One of the well-known characteristics of the Down syndrome is that its incidence increases greatly with the advancing age of the mother (this tendency also is apparent in the Klinefelter syndrome): the incidence of the Down syndrome in mothers under 30 years of age is approximately one in 2,000–3,000 births, whereas in mothers over 40 it rises to about one in 40 births.[8] The reason for this age correlation is unknown, although, as stated by German,[8] "Numerous ideas have been advanced to explain chromosomal nondisjunction in older mothers, including infection of the oocyte with a microorganism, an abnormal effect of the nucleolus in a cell held in prophase for so many years, autoimmunity, radiation effect, deterioration of the mitotic apparatus in the oocyte, decreased numbers of synapses in eggs released late in life, and delayed release or fertilization of an ovum." It is well known that chromosome damage may be caused by radiation (x-rays, gamma rays), drugs and chemicals (LSD, caffeine, sulfur dioxide), viruses and even by certain physical forces such as ultrasonic vibrations.

An interesting corollary to the discussion of the Down syndrome is the frequency with which chromosomal aberrations of one sort or another appear in overall births (both live births and spontaneous abortions) in the general population. Although figures may vary according to different sources, the following is probably a quite accurate representation of this phenomenon. It has been reported that one in every 250 live-born children presents some chromosomal abnormality. Certain studies indicate that in spontaneous abortions during the first trimester of pregnancy, chromosomal aberrations are apparent in greater than one in five instances and that this figure is more than two in three of very young aborted embryos. The conclusion may be drawn that chromosomal abnormalities are a very prominent feature in embryonic wastage.[8] In addition, abnormal chromosomes are very prominent in neoplasms (cancers), although it is not known whether this relationship is primary (causative) or secondary (resultative) in nature. A constant feature of chronic myelogenous leukemia is the presence of a small, abnormal chromosome known as the Philadelphia or Ph chromosome.

The technician or physician who prepares and analyzes the karyotype of a given individual must be extremely careful to examine the chromosomal complement for *structure* as well as for *number*. For example, through various *translocations*, i.e., the transposition of a piece of a chromosome or an entire chromosome to another, a given complement may contain 46 chromosomes and yet be abnormal or contain 45 or even 44 chromosomes without any associated phenotypic defects. An example of this is presented in Figure 10–4, B, in which the chromosome number is the normal 46 but in which one small acrocentric autosome (probably a 21) is translocated onto a 14, resulting in effect in trisomy 21 (the Down syndrome). Conversely, it is apparent that a translocation of this sort in a normal individual would result in a karyotype with only 45 chromosomes and that two such translocations would result in a total of 44 chromosomes, and so on.

Due to the tremendous advances made in human cytogenetics over the past two and a half decades, physicians and other scientists are currently in a position to effectively reduce the incidence of the Down syndrome and related disorders. Through amniocentesis (the withdrawal of amniotic fluid containing sloughed-off somatic cells from the fetus), the physician is able to detect if a chromosomal abnormality is present. If so, a therapeutic abortion may be indicated. Genetic counseling may be used when parents have already given birth to one Down child or when there is evidence that the prospective mother may be the carrier of a translocation that predisposes to mongolism or some other cytogenetic disorder. In such instances the prospective parents are encouraged to weigh the odds of producing normal vs. abnormal children and to make their decision on the basis of this consideration.

CASE 10–5: BREAST CANCER

History

A 56-year-old woman noticed a dimpling of the skin of the right breast about 3 cm superior to the areola and in a midsagittal plane, i.e., one that extended through the nipple. Closer inspection revealed that the skin was firmer and tauter than ordinary (it actually felt hard to the touch), and it had the "dimpled" appearance of an orange peel or pigskin leather. A visit to her physician confirmed her suspicions and fears that she might have a breast tumor. The doctor, after manual and thermographic examinations, established that the tumor was approximately 2–3 cm in diameter and that it was placed quite deeply in the breast, close to the anterior surface of the pectoralis major muscle. A biopsy demonstrated malignancy. Manual examination of the patient's right axilla did not demonstrate any detectably enlarged lymph nodes. The woman was scheduled for surgery the following morning, at which time a radical mastectomy was performed. A pathologic examination revealed a slight degree of metastases in about 25% of the axillary nodes removed during the operation. The surgeons believed that all cancerous tissue had been removed, and the general consensus was that the prognosis for total recovery and freedom from future metastatic sites was good. Indeed, 7 years after the operation, the woman was free from any detectable signs of cancer.

Questions

Discuss the lymphatic drainage of the breast, giving specific consideration to the draining nodes of the various regions of this structure.

What was the reason for the dimpling of the skin of this woman's breast?

Although you are not expected to have any detailed surgical

knowledge at this point, attempt to describe anatomically the radical mastectomy procedure.

Do you feel that this procedure is justified in the majority of cases of breast cancer or would simple mastectomy, combined with axillary lymph node dissection, or simple mastectomy alone, suffice?

Discussion

A knowledge of the lymphatic drainage of the *breast* (Fig. 10–5) is exceedingly important due to the frequency of transmission through these channels of metastatic malignant cells. Basically, the *central portion of gland,* the skin over this area and the areola and nipple drain eventually to two trunks, which as efferent vessels enter the *pectoral group of axillary lymph nodes.* It is important to note that, although this is the pathway most commonly followed, lymph from various portions of the breast may drain directly to any or all of the groups of axillary nodes (for a discussion of the lymph nodes of the axilla, see Case 1–1). Lymph from the above-mentioned areas of the breast also may drain to inferior deep cervical nodes,

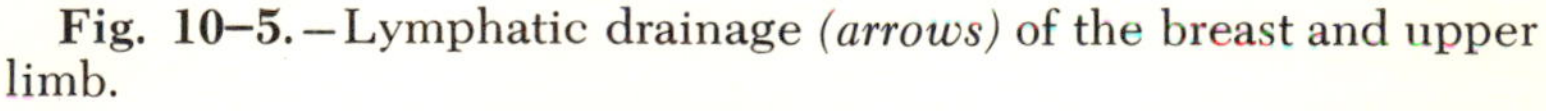

Fig. 10–5.—Lymphatic drainage *(arrows)* of the breast and upper limb.

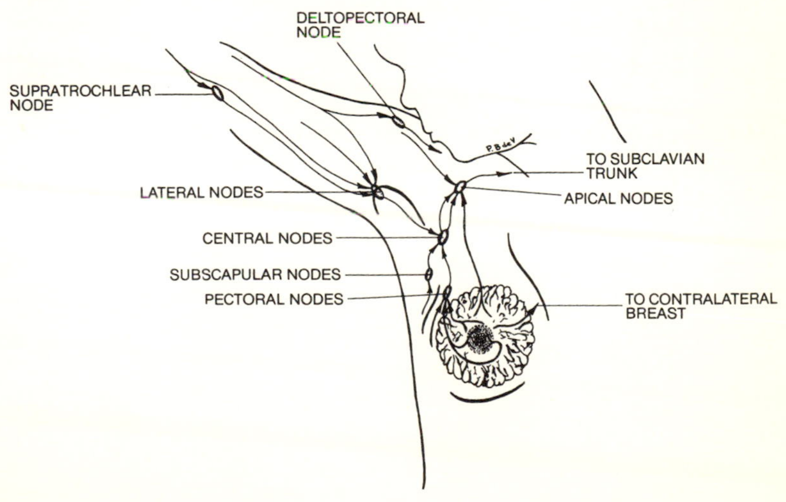

deltopectoral nodes or parasternal (sternal, internal thoracic) nodes bilaterally. Efferent lymphatic vessels from the medial aspects of the breast follow the course of perforating blood vessels through the pectoralis major and intercostal muscles to *parasternal nodes*. The latter are small (1–2 cm in diameter) and few in number (3–5 on a side). Lymph from the medial aspect also may cross the midline to parasternal nodes and even to those of the contralateral axilla, which probably explains the occasional observation of bilateral axillary metastases from a tumor in only one breast. Lymph from the inferomedial aspect of the breast may drain to the anterior abdominal wall (a plexus on the rectus sheath) and from there to subphrenic and subperitoneal plexuses.

According to Cutler,[9] "Enlargement of the axillary lymph nodes does not necessarily indicate metastasis, nor does the absence of enlarged nodes eliminate the possibility of the existence of metastatic deposits." Nevertheless, the axillary lymph nodes are the commonest site of metastasis from tumors of the breast, and their enlargement in a female patient should at least arouse suspicion in the physician as to the possibility of the presence of breast cancer. Reports of the incidence of axillary metastasis in specimens of breasts amputated for cancer vary from 56–79%.[9] Interestingly, there may be no axillary metastases in very large carcinomas of the breast and, conversely, there may be axillary involvement from minute mammary tumors. In other words, the degree of malignancy in certain cases is inversely proportional to the size of the tumor. Metastasis to the anterior mediastinal lymph nodes is rare in the very early stages of breast cancer. On the other hand, the internal thoracic (internal mammary) lymph nodes are involved in approximately one third of presumably operable mammary cancers. Metastasis to these lymph nodes is much more common when the primary tumor is located in the medial half, rather than in the lateral half, of the breast (approximately 50% of medially placed tumors metastasize to internal thoracic nodes). Another interesting and important statistic is that in almost half of the cases in which axillary nodes are already involved, mediastinal nodes also demonstrate metastasis, no matter where the initial site was located.[9] Urban[10] reported that 58% of patients with axillary node involvement also demonstrated spread to the internal thoracic chain. Metastases to the latter nodes were found most frequently in the second interspace and in decreasing frequency in the third, first, fourth and fifth interspaces. Urban also reported that the upper medial quadrant of the breast is the least likely to demonstrate over-all metastases to both internal thoracic and axillary nodes. The tumor in this specific case was located directly above the nipple, an area that projects to the upper medial quadrant. In addition, Urban noted that statistically 45% of breast tumors are found in the upper lateral quadrant, 15% in the upper medial, 10% in the lower lateral, 5% in the lower medial and 25% in the area of the nipple and areola. This observation of by far the greatest frequency of breast cancer in the upper lateral quadrant probably explains why axillary node metastases are found much more often than internal thoracic metastases when mammary tumors are described in general terms (i.e., without regard to specific site of origin).

The dimpling of the skin observed in this patient, which is frequently a diagnostic symptom of an underlying breast tumor, is caused by pressure exerted by the mass on connective tissue septa that extend from the base of the breast to the skin. The result in effect is a bowing and therefore a *shortening* of the septa, which results in an *impression* or *depression* of the skin rather than the usually observed and expected "lump" characteristic of a tumor. The student is encouraged at this point to review both the gross and microscopic anatomy of the human breast.

Development of the radical mastectomy operation as it is known today, with, of course, numerous individual variations, is credited to Halsted[11] in 1907. The operation consists basically of making a large, vertically oriented, elliptical incision that encompasses the affected breast. The pectoralis major muscle is located and divided near its insertion on the greater tubercle of the humerus. The pectoral branches of the thoracoacromial vessels are ligated, and the vessels plus the medial and lateral pectoral nerves are then cut. The breast and underlying pectoralis major muscle are reflected laterally to expose the pectoralis minor muscle. Its lateral edge is freed from the deep fascia, and the muscle is then elevated and divided near its insertion on the coracoid process. The blood vessels to this muscle are ligated and cut, and any remaining medial and/or lateral pectoral nerve branches are similarly incised. Care is taken throughout the operation to "avoid all the means by which malignant cells may be transplanted into the wound or pressed from accessible parts to others that are inaccessible."[9] The pectoralis minor muscle is removed, and the remaining branches of the axillary artery that are in the field of operation (highest or superior thoracic artery, certain thoracoacromial branches, lateral thoracic artery) and branches of the axillary vein are ligated and cut. The breast and pectoralis major muscle are removed. The fat, fascia and lymph nodes of the axilla and pectoral region are dissected and removed *en masse.* The fat and fascia also are dissected from the rectus abdominis muscle anteriorly and from the serratus anterior muscle laterally. It should be noted that the subscapular vessels and the thoracodorsal and long thoracic nerves are left intact. Final cauterization of any bleeding vessels is performed and the skin incision is closed. Patients so treated are permitted to get out of bed on the second day and to leave the hospital anywhere from the fifth to the tenth days.[9]

Controversy exists and will no doubt continue to exist with regard to the value of the radical mastectomy procedure compared to the simple operation combined with radiation or with axillary lymph node dissection. The former is obviously more debilitating (since the pectoral muscles are removed) and more disfiguring, and modern attitudes are beginning to doubt the efficacy of the removal of these muscles. Certainly, the high degree of axillary lymph node metastases warrants a detailed examination of these structures in most, if not all, cases of breast cancer. Although such considerations are in the strict sense outside the scope of this book, you, as future physicians, should be cognizant of the psychologic trauma incurred by removal of a breast alone, not to mention that which results from a reduced functional ability of the upper limb.

CASE 10–6: GUNSHOT WOUND OF THE SPINAL CORD

History

A 22-year-old man was accidentally shot in the back with an "unloaded" .22 caliber rifle following a target practice session with a friend. The bullet entered the back in the midsagittal plane at the level of the spine of the first lumbar vertebra. It smashed the spinous process, passed through the vertebral canal and lodged in the inferior aspect of the body of the vertebra. The young man was raced to a local hospital where the bullet was removed surgically. Immediately following surgery the patient complained of or demonstrated upon examination the following symptoms or disorders: (1) flaccid paralysis of most of the muscles of the pelvis and perineum, including the levator ani, external anal sphincter, sphincter urethrae, superficial and deep transverse perineal muscles and ischiocavernosus and bulbocavernosus muscles; (2) loss of sensation in most of the perineum and down the posterior aspect of the thigh and calf as far as the foot; (3) paralysis of the bladder, resulting in distention without automatic emptying but, rather, with a passive dribbling of urine; (4) paralysis of the sigmoid colon and rectum, resulting in an absence of spontaneous defecation, accompanied, however, by a degree of fecal incontinence and (5) loss of the capacity for erection and ejaculation. In addition, there was some loss in the motor activity in the gluteal region, in the posterior aspects of both thighs and in the anterior and posterior aspects of both legs. There was no motor loss in the muscles that flex the hip or extend the knee. Although there was partial recovery of the patient's ability to control defecation, the other dysfunctions were permanent.

Questions

What specific portion of what single structure did the bullet damage in order to produce the collective symptoms described in this case?

What named nerves were affected by this injury, and what losses (motor, sensory and/or autonomic) resulted from their central interruption?

Review the autonomic nerve supply to the organs of the pelvis and perineum, giving special attention to the innervation of the bladder as it relates to micturition and to the penis as it relates to erection.

Would you expect similar observations with regard to motor

and sensory losses had the wound been two vertebral levels lower?

Discussion

The specific structure injured by the bullet was the *conus medullaris* of the spinal cord; the resulting symptoms have been referred to as the *"conus syndrome."*[12] Figure 10–6 demonstrates that the tapering end of the spinal cord terminates at the middle of the body of L. V. 2 in this case. Below this point the spinal nerves course inferiorly as the *cauda equina* ("horse's tail") to exit at their respective intervertebral

Fig. 10–6.—Relationship of the lower *spinal cord* and *cauda equina* to the inferior portion of the vertebral canal (based on Crosby *et al.*[13] and other sources).

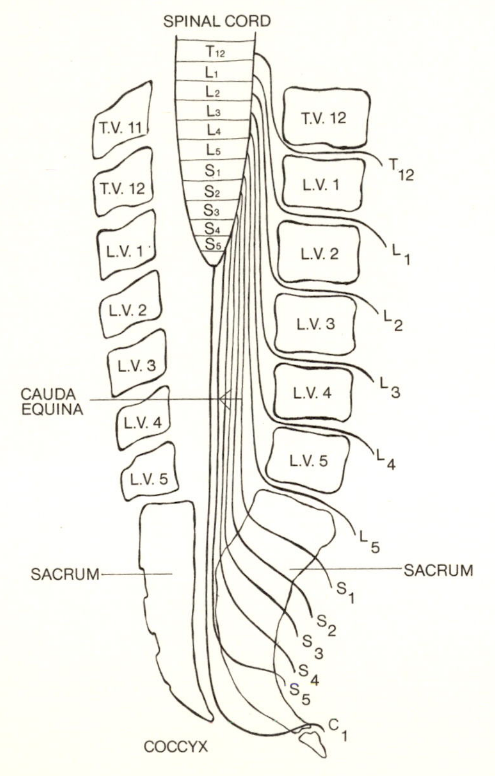

or sacral foramina. The conus medullaris contains basically the sacral and coccygeal spinal cord segments. Further reference to Fig. 10–6 demonstrates that a bullet that passed through the spinous process of L. V.1 and lodged in the inferior aspect of the body of this vertebra, as occurred in this case, would probably spare the first sacral cord segment. Although it is difficult to predict what secondary damage might be caused by splintering bone fragments, it was concluded in the present case by neurologic examination that S_1 was not injured.

Following is a list of nerves that contain fibers from any of the sacral segments two to five or from the coccygeal segment of the cord. In each instance a brief discussion of the distribution and function is included; these functions would be lost if the nerve contained fibers only from S_2 or below, or modified (i.e., partial loss) if the nerve contained higher fibers as well. The numbers in parentheses after each nerve represent its total components. The student is advised at this point to carefully study the formation of the sacral plexus and the contributions to it from the lower lumbar nerves.

1. The *inferior gluteal nerve* (L_5; $S_{1\text{-}2}$) leaves the pelvis through the greater sciatic foramen below the piriformis muscle to supply the gluteus maximus muscle. The latter extends the thigh or, if the thigh is fixed, extends the trunk or pelvis.
2. The *nerve to the obturator internus* (L_5; $S_{1\text{-}2}$) also leaves the pelvis through the greater sciatic foramen below the piriformis. It gives a branch to the superior gemellus, crosses the ischial spine, re-enters the pelvis through the lesser sciatic foramen and ends in the obturator internus muscle. These two muscles are lateral rotators of the thigh; they also tend to stabilize the hip. The obturator internus also is a weak abductor of the thigh.
3. The *posterior femoral cutaneous nerve* ($S_{1\text{-}3}$) leaves the pelvis with the above two nerves. It gives off *inferior cluneal nerves,* which wind around the lower border of the gluteus maximus muscle to supply the skin of the buttock. It also gives off a *perineal branch* (or branches), which courses anteriorly at the superomedial angle of the thigh to supply the perineum and the skin of the external genitalia. The main body of the

posterior femoral cutaneous nerve courses inferiorly to supply the posteromedial aspect of thigh and calf. Loss of sensation in these areas gives the appearance of the so-called saddle anesthesia.[12]

4. The *perforating cutaneous nerve* (S_{2-3}) pierces the sacrotuberous ligament and supplies a small area of skin over this structure.

5. The *sciatic nerve* (L_{4-5}; S_{1-3}) also leaves the pelvis through the greater sciatic notch below the piriformis muscle. In some instances it leaves as two separate entities, one often piercing the piriformis. Usually, however, the sciatic divides into the *tibial* and *common peroneal nerves* in the posterior thigh. Direct branches from the sciatic nerve supply the hamstring muscles, which collectively extend the thigh and flex the leg. The tibial and common peroneal nerves supply the muscles of the leg and foot and most of the skin of these areas.

6. The *nerve to the piriformis* (S_{1-2}) enters that muscle directly from within the pelvis. The piriformis rotates the thigh laterally, stabilizes the hip and is a weak abductor of the thigh.

7. The *nerves to the levator ani* and *coccygeus* muscle (S_{3-4}) enter these muscles from their pelvic aspects. These two muscles, which constitute the pelvic diaphragm, support the pelvic viscera and assist in maintenance of increased intra-abdominal pressure upon contraction of the abdominal wall musculature.

8. The *pudendal nerve* (S_{2-4}) (representing one of the most severe losses emanating from the injury in this patient) exits from the pelvis through the greater sciatic foramen below the piriformis, crosses the ischial spine, re-enters the pelvis through the lesser sciatic foramen and then courses anteriorly toward the urogenital triangle in the pudendal canal. It gives off the *inferior rectal nerve,* which traverses the ischiorectal fossa as many branches and supplies the sphincter ani externus muscle, the anal canal up to the pectinate line and the skin around the anus. Loss of motor fibers to the sphincter ani externus accounted for the fecal incontinence in this patient. The pudendal nerve then divides into the *perineal nerve* and the *dorsal nerve of the penis*. The former is motor to the bulbo-

cavernosus, ischiocavernosus and superficial transverse perineal muscles of the urogenital triangle and is sensory to the scrotum via the *posterior scrotal nerves* (*posterior labial* in the female). It is important to note that sensation was not lost from the anterior portion of the scrotum in this patient since it is supplied by anterior scrotal branches (anterior labial in the female) of the ilioinguinal nerve (L_1). The *dorsal nerve of the penis* (*clitoris* in the female) is motor to the sphincter urethrae (partially responsible for the dribbling of urine) and the deep transverse perineal muscles. It is sensory to the skin, prepuce and glans of the penis.

9. The *pelvic splanchnic nerves* or *nervi erigentes* (S_{2-4}) represent the sacral portion of the craniosacral outflow or parasympathetic division of the autonomic nervous system (see Case 8–1 for a general discussion of this system). It is probably safe to say that it is the loss of innervation normally carried by these nerves that is most distressing to the patient with the conus syndrome. The cell bodies of preganglionic parasympathetic neurons lie in the intermediolateral area of sacral spinal cord segments two to four. This cell column is not as clearly delineated as that which contains the preganglionic sympathetic cell bodies in the thoracolumbar portion of the spinal cord. The preganglionic parasympathetic neurons enter the corresponding sacral nerves and then leave them as visceral branches that take part in the formation of the *pelvic (inferior hypogastric) plexus*. This plexus lies bilaterally in the tough connective tissue at the sides of the pelvic organs. Branches of the plexus contain preganglionic neurons that synapse upon cell bodies of postganglionic neurons in or near the walls of the descending colon, sigmoid colon, rectum, bladder, prostate, seminal vesicle, testis (uterus, uterine tubes and ovaries in the female), erectile tissue and blood vessels supplying the pelvic organs. Parasympathetic stimulation results in contraction of the smooth muscle of the distal portion of the large intestine and of the detrusor muscle of the bladder (see below). In addition, these nerves are responsible for stimulation of glands, arterial vasodilation and erection of cavernous tissue in both the male and female.

10. The *coccygeal plexus* (usually S_{4-5} and the coccygeal

nerve) supplies the sacrococcygeal joint, the coccyx and the overlying skin of the area.

As stated by Brodal,[12] "The physiology of micturition is a complex subject, and the nervous factors involved are not yet completely known." It was formerly believed that parasympathetic stimulation of the detrusor muscle (the main muscle mass of the bladder) occurred concomitantly with sympathetic inhibition of the musculature of the internal sphincter. Crosby *et al.*[13] suggested that there is no true internal sphincter that relaxes under sympathetic influence during micturition but, rather, simply a continuation of the detrusor muscle in longitudinally running bands extending onto the urethra. This seems to be supported by the observation of "dribbling urination" in patients with the conus syndrome. In these cases sympathetic innervation of the bladder remains intact, but spontaneous voiding of urine, which would be expected upon relaxation of a true internal sphincter, does not take place. This seems to suggest that upon parasympathetic denervation the bladder fills to the point where it simply "overflows."

The autonomic innervation to the genitalia is similarly not well understood. Sympathetic fibers to seminal vesicle, vas deferens and prostate are vasoconstrictor and motor in nature. Although ejaculation is believed to result from sympathetic stimulation (this phenomenon has been demonstrated experimentally), loss of parasympathetic innervation only, as occurs in the conus syndrome, inhibits or eliminates ejaculation. This may be related to the loss of the capacity for erection as well as to the loss of afferents from most of the genitalia in conus injuries. According to Brodal,[12] " . . . the main effect of the parasympathetic innervation is to be found in the vasodilation of the erectile tissues of the genital organs, which follows the stimulation of these nerves." It is this function that earned them the name of nervi erigentes. Many anatomists also believe that the ischiocavernosus and bulbocavernosus muscles, upon contraction, help to maintain erection by compressing the crura and bulb, respectively, thereby retarding the return of venous blood from the penis.

Damage to the cauda equina (e.g., two vertebral levels inferior to the conus medullaris) will result in symptoms that are similar, perhaps even indistinguishable, from the conus syndrome. However, due to the fact that the cauda equina contains lumbar as well as sacral nerves, lesions caused by traumatic injury or tumors in this area might be expected to result in sensory and motor losses in nerves originating from higher levels of the cord. Indeed, because some nerves may be spared (e.g., pushed aside) in such conditions, the resultant picture might be one in which there are losses in some but not all of both lumbar and sacral nerves. We might also expect to observe discrepancies in bilateral losses in such cases.

CASE 10–7: UTERINE PROLAPSE

History

A 60-year-old multiparous woman presented with low back pain and a mass protruding from the vagina. The latter became especially prominent during straining, i.e., with increased intra-abdominal pressure. A history revealed that she had given birth to five normal children over a 9-year period. The first two deliveries were performed without episiotomy, and she was told that the fourth birth was difficult, requiring extensive use of forceps. The last birth occurred when she was 34 years of age. A pelvic examination revealed a cystocele (protrusion of the urinary bladder through the wall of the vagina), rectocele (in this case, part of the rectum bulged into the vaginal wall) and a partially prolapsed uterus. She was treated surgically, which eliminated the problem of the protruding viscera, and this, combined with weight loss through dieting, alleviated much of the back pain.

Questions

Define the boundaries of the pelvis.

Describe the anatomy of the pelvic diaphragm and the other mechanisms of support for the pelvic viscera.

What are the probable causes of this general condition and how might it be prevented?

Discussion

The human *pelvis* (Latin = "basin") is that portion of the trunk inferior to the abdomen and separated from the perineum by the pelvic diaphragm (see also Case 10–1). It may be divided for descriptive purposes into a *pelvis major (greater* or *false pelvis)* and a *pelvis minor (lesser* or *true pelvis),* which are separated by the *pelvic brim.* The latter is a demarcation consisting of the pubic crests, pectineal lines, arcuate lines of the ilia bilaterally and the anterior margin of the base of the sacrum in the posterior midline. The pelvis major, superior to the pelvic brim, basically consists of the iliac fossae and is best considered as a subdivision of the abdominal cavity. Thus, when we speak of the pelvis, we are referring to the pelvis minor, or that area of the body between the pelvic brim and the pelvic diaphragm.

The *pelvic diaphragm* (see Fig. 10–1) is made up of the bilateral *levator ani* and *coccygeus muscles.* It is perforated by the anal canal and urethra in both sexes and additionally by the vagina in the female. The levator ani arises from the inner surface of the pubis; from the *arcus tendineus,* a thickened band of parietal pelvic (obturator internus) fascia; and from the ischial spine. It is subdivided into three muscles, the *puborectalis, pubococcygeus* and *iliococcygeus,* their respective origins and insertions being indicated by their names. The puborectalis forms a sling around the anorectal junction (Fig. 10–7); where it comes in contact with the prostate gland in the male and with the vagina in the female, its fibers are

Fig. 10–7.—The *puborectalis muscle,* which forms a *sling* around the lower portion of the *rectum.*

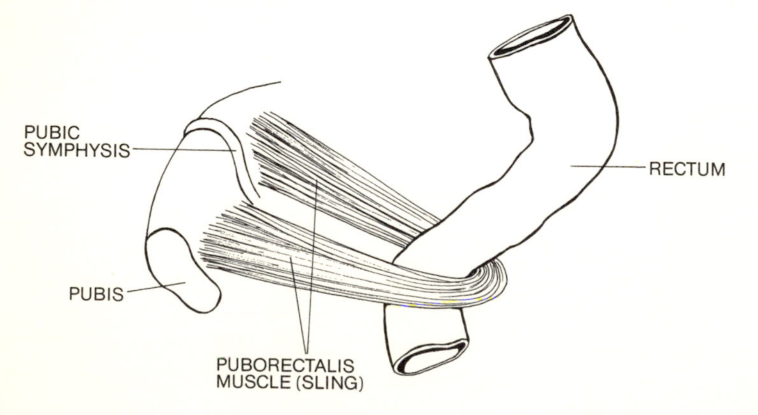

referred to as the *levator prostatae* and *pubovaginalis,* respectively. The puborectalis is important in maintaining the position and integrity of the pelvic viscera. The *coccygeus (ischiococcygeus)* lies posterior to the levator ani, passing from the ischial spines to the lateral margins of the coccyx. The coccygeus may be largely tendinous in nature. The pelvic diaphragm is supplied by S_3 and S_4; it supports the pelvic viscera and assists in increasing intra-abdominal pressure.

The *pelvic (endopelvic) fascia* is also very important in maintaining the integrity of the floor of the pelvis and in preventing displacement of the pelvic viscera. It usually is subdivided for descriptive purposes into parietal and visceral portions. The *parietal pelvic fascia* lines the internal walls of the pelvis except where it is interrupted by bone, in which case it is replaced by periosteum. It is continuous superiorly with the extraperitoneal fascia of the abdominal cavity (the endoabdominal fascia, e.g., the transversalis fascia, iliac fascia). The parietal pelvic fascia lines the obturator internus muscle and the superior and inferior surfaces of the pelvic diaphragm. Technically speaking, the inferior fascia of the pelvic diaphragm and the obturator internus fascia inferior to the arcus tendineus are in the perineum rather than in the pelvis. The pudendal canal (see Case 10–1) lies within an envelope of parietal pelvic fascia.

The *visceral pelvic fascia* forms a packing for the pelvic viscera and also ensheaths the blood vessels of the pelvis. In several places the visceral fascia is condensed into the so-called "ligaments" of the pelvis that assist in support of the pelvic organs. Basically, these are (1) the *puboprostatic ligaments* in the male and their counterpart in the female and the *pubovesical ligaments,* which occupy the medial edges of the puborectalis muscles and provide support for the bladder; (2) the *lateral cervical (cardinal) ligaments,* which extend between the ischial spines and the lateral aspects of the uterine cervix and the rectum (these serve as important supports for these structures); and (3) the *uterosacral ligament,* which extends between the middle of the sacrum and the cervix.

The condition experienced by this patient is usually referred to as *relaxed pelvic floor* or *prolapse* of one or more pelvic organs. It is generally accepted that it is associated with a number of conditions and that no single cause usually can be held responsible.[14] Turell[14] cites such probable causes as weakened or attenuated pelvic fascia, weakness of sphincters and levators, chronic straining and neurologic disorders. Repeated tearing and subsequent scar formation of the perineal body (Case 10–1) are no doubt related to these conditions. Stoddard[15] states, "Young tissues can be stretched considerably and still maintain adequate support. In postmenopausal patients the pelvic tissues atrophy, and this, coupled with damage years earlier and/or heavy work, usually results in symptom-producing relaxations." He adds, "We have little or no control over the quality of tissues and the size of babies. By performing timely episiotomy, undue stretching of perineal and bladder supports may be decreased. Exercise of the pubococcygeus muscles may help to maintain tissue integrity to some extent."

REFERENCES

1. Tenney, B., and Little, B.: *Clinical Obstetrics* (Philadelphia/London: W. B. Saunders Co., 1961).
2. Barr, M. L., and Bertram, E. G.: A morphological distinction between neurons of male and female, and the behaviour of the nucleolar satellite during accelerated nucleoprotein synthesis, Nature 163:676, 1949.
3. Hsu, T. C., and Pomerat, C. M.: Mammalian chromosomes in vitro. II. A method for spreading the chromosomes of cells in culture, J. Hered. 44:23, 1953.
4. Tjio, J. H., and Levan, A.: The chromosome number of man, Hereditas 42:1, 1956.
5. Nowell, P. C.: Phytohemagglutinin: An initiator of mitosis in cultures of normal human lymphocytes, Cancer Res. 20:462, 1960.
6. Lyon, M. F.: Gene action in the X-chromosome of the mouse (*Mus musculus* L.), Nature 190:372, 1961.
7. Bartalos, M., and Baramki, T. A.: *Medical Cytogenetics* (Baltimore: Williams & Wilkins Co., 1967).
8. German, J. L., III: Studying human chromosomes today, Am. Sci. 58:182, 1970.
9. Cutler, M.: *Tumors of the Breast* (Philadelphia/Montreal: J. B. Lippincott Co., 1962).
10. Urban, J. A.: Clinical experience and results of excision of the

internal mammary lymph node chain in primary operable breast cancer, Cancer 12:14, 1959.
11. Halsted, W. S.: The results of radical operations for the cure of cancer of the breast, Ann. Surg. 46:1, 1907.
12. Brodal, A.: *Neurological Anatomy* (2d ed.; New York/London/ Toronto: Oxford University Press, 1969).
13. Crosby, E. C., *et al.*: *Correlative Anatomy of the Nervous System* (New York: MacMillan Co., 1962).
14. Turell, R.: *Diseases of the Colon and Anorectum* (Philadelphia/ London/Toronto: W. B. Saunders Co., 1969).
15. Stoddard, F. J.: *Case Studies in Obstetrics and Gynecology* (Philadelphia/London: W. B. Saunders Co., 1964).

INDEX

C

F

I

J

M

Q

R